高职高专"十三五"规划教材

化工专业英语

谢承佳　张培培　主编
张崎静　主审

化学工业出版社

·北京·

《化工专业英语》是为了提高化工技术类专业学生化工英语信息获取能力和应用能力，培养与职业能力结构要求相一致的高素质技能型人才而编写的。本教材立足于化工技术专业典型工作岗位群（开发员、操作工、安全员、检验工、环保员、销售员）建立六个典型工作情境，分别是 Product Development 产品开发、Mass Production 批量生产、Chemical Engineering Safety 化工安全、Product Testing 产品检测、Environment Protection 环境保护、Marketing & Selling 产品销售。每个情境分为学习目标、情境导入、精读、练习、语法、拓展、阅读材料等。编者从化工行业典型工作过程及职业岗位来组织内容，材料来源于英文原版书籍、杂志和英文化工方面的相关网站，由化工专业教师、英语教师与行业企业合作，突出专业英语"专业+语言"的特征，借助英语工具，开展化工专业实践，强化应用，紧密结合化工生产实际。

《化工专业英语》图文并茂、操作性强、覆盖面广、难度适中，可作为应用化工、石油化工、精细化工、环境保护、工业分析与检验等高职高专化工类专业的英语教材，也可作为从事化工产品生产工作人员的参考书。

图书在版编目（CIP）数据

化工专业英语/谢承佳，张培培主编．—北京：化学工业出版社，2018.3（2021.8重印）
ISBN 978-7-122-31472-7

Ⅰ.①化… Ⅱ.①谢… ②张… Ⅲ.①化学工业-英语-教材 Ⅳ.①TQ

中国版本图书馆CIP数据核字（2018）第020290号

责任编辑：刘心怡　蔡洪伟　　　　　　　　　　装帧设计：关　飞
责任校对：王　静

出版发行：化学工业出版社（北京市东城区青年湖南街13号　邮政编码100011）
装　　装：北京七彩京通数码快印有限公司
787mm×1092mm　1/16　印张 8¾　字数244千字　2021年8月北京第1版第3次印刷

购书咨询：010-64518888　　　　　　　　售后服务：010-64518899
网　　址：http://www.cip.com.cn
凡购买本书，如有缺损质量问题，本社销售中心负责调换。

定　价：29.00元　　　　　　　　　　　　　　　　　版权所有　违者必究

前　言

化工专业英语课程是高职院校英语教学的重要组成部分，本门课程可以提高化工技术类专业学生化工英语信息获取能力和信息产出能力，使学生具备从事化学化工行业必需的英语素质和职业素养，培养与职业能力结构要求相一致的高素质技能型人才。

本教材以《高等职业教育英语课程教学要求》为建设思路，根据专业设置和学生未来就业的行业及岗位（群）需求制定体现职业教育教学理念和教学模式的行业英语课程，以化工行业企业英语职业能力需求为依据，以职业能力培养为重点，以典型化工产品生产过程为主线建立典型工作情境。

本教材基于地方企业调研，分析化工行业英语职业能力需求，对完成化工行业各岗位需要工作任务应具备的职业能力做出详细的描述，同时对工作任务、职业能力按逻辑关系进行排序，立足化工技术专业典型工作岗位群（开发员、操作工、安全员、检验工、环保员、销售员）建立六个典型工作情境，即 Product Development 产品开发、Mass Production 批量生产、Chemical Engineering Safety 化工安全、Product Testing 产品检测、Environment Protection 环境保护、Marketing & Selling 产品销售。在每一情境中，分为以下几部分。

① Objectives　提供本情境教学完成后学生应掌握的主要职业技能及语言技能。
② Warming-up　情境的导入部分，通过图文并茂的形式引入情境主体。
③ Text　精读部分，选择的课文与情境目标相关度高，可根据具体教学课时选择内容。
④ Comprehension　以选择题、填空题、问答题的形式考查学生对课文的理解。
⑤ Vocabulary building　选择适量的高频词汇及核心专业词汇培养学生的语言能力。
⑥ Exercise　通过练习巩固学生对词汇的掌握。
⑦ Extension　选择了一些与精读文章相关度较高的小知识培养学生的知识迁移能力。
⑧ Reading material　选择了一些难度较高的文章满足差异性教学，也可作为泛读课文。
⑨ Supplementary knowledge　介绍了科技文献、专利、说明书、操作手册、合同等常见文体，培养学生的职业能力。

本教材为校企开发教材。情境 1 由扬州工业职业技术学院谢承佳和南京炼油厂的朱宇清高工共同编写，情境 2 由扬州工业职业技术学院张培培编写，情境 3 由扬州工业职业技术学院陈秀清编写，情境 4 由扬州工业职业技术学院马振雄和谢承佳共同编写，情境 5 由谢承佳和南京炼油厂的谢承芳高工共同编写，情境 6 由扬州工业职业技术学院钱婧编写，全书由扬州工业职业技术学院张崎静主审。

由于笔者水平有限，疏漏在所难免，希望读者不吝赐教，批评指正，以便再版时更正和改进。

编者
2017 年 7 月

Contents

Product Development 1

Warming-up What Are Substances Composed of? 2
Text A Elements and Compounds 3
Text B Systematic Nomenclature of Binary Compounds 7
Text C Properties of Aqueous Ammonia 11
Reading material The Haber Process for Ammonia Synthesis 15
Supplementary knowledge Structure of Patent Documents 17

Mass Production 21

Warming-up Development of Chemical Engineering 22
Text A Types of Chemical Reactors 23
Text B Basic Distillation Equipment and Operation 27
Text C Petrochemicals 33
Reading material Ethylene Production 38
Supplementary knowledge Structure of Equipment Manual 40

Chemical Engineering Safety 43

Warming-up What Do These Public Signs Mean? 44
Text A General rules for classification and hazard communication of chemicals 44
Text B Technical & Safety Instructions for Chemicals 48
Text C Accident Investigation Report 53
Reading material How to Use a Fire Extinguisher? 59
Supplementary knowledge Structure of Material Safety Data Sheet 64

Product Testing 69

Warming-up What Are the Usages of the Following Instruments? 70

Text A	Instruments	71
Text B	Redox Titration	75
Text C	Acid-Base Indicators	79

Reading material Gas and Liquid Chromatography 82

Supplementary knowledge Structure of Operating Manual 84

Environment Protection 89

Warming-up Types of Pollution 90

Text A	Water Pollution	92
Text B	Biological Oxygen Demand Monitoring	97
Text C	Recycling and Reuse	101

Reading material Air Pollution 105

Supplementary knowledge Structure of a Journal-Style Scientific Paper 107

Marketing & Selling 111

Warming-up Which Chemical Firm will be Your Target? Why? 112

Text A	BASF and Other Top Chemical Firms	113
Text B	Inquires	117
Text C	Sale & Purchase Contract for BRAZIL Iron Ore Fines	121

Reading material 127

Supplementary knowledge Samples of Some International Trade Documents 130

Product Development

Objectives:

After finishing this module, you are able to:
- Name chemicals properly
- Find useful information of chemicals quickly by using chemical handbooks
- Confirm useful information in the patent

Warming-up

What Are Substances Composed of?

The theory of four elements

As recently as a few thousand years ago, western scientists considered the whole earth to be made of 4 elements: Earth, Air, Fire, and Water. Air was the underlined element: this single substance made up everything in the world.

Discovery of oxygen

In 1774, an English cleric named Joseph Priestley observed an interesting phenomenon: when mercuric tox is heated to certain temperature, a colorless gas (later renamed oxygen by Antoine Lavoisier) and a silvery liquid metal were produced.

Atomic theory

ELEMENTS			
	Wt.		Wt.
⊙ Hydrogen	1	© Copper	56
⊕ Azote	5	Ⓛ Lead	90
● Carbon	6	Ⓢ Silver	190
○ Oxygen	7	Ⓖ Gold	190
⊗ Phosphorus	9	Ⓟ Platina	190
⊕ Sulfur	13	Ⓜ Mercury	167

It was John Dalton, in the early 1800s, who determined that each chemical element was composed of a unique type of atom, and that atoms differed by their masses. He devised a system of chemical symbols and, having ascertained the relative weights of atoms, arranged them into a table.

Atoms combine into molecules

Italian chemist Amedeo Avogadro found that the atoms in elements were combined to form molecules. Avogadro proposed that equal volumes of gases under equal conditions of temperature and pressure contain equal numbers of molecules.

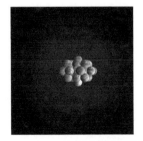

Text A Elements and Compounds

Elements are pure substances that cannot be decomposed into simpler substances by ordinary chemical changes. At present there are 109 known elements. Some common elements that are familiar to you are carbon, oxygen, aluminum, iron, copper, nitrogen, and gold.

About 85% of the elements can be found in nature, usually combined with other elements in minerals and vegetable matter or in substances like water and carbon dioxide. Copper, silver, gold, and about 20 other elements can be found in highly pure forms. Sixteen elements are not found in nature; they have been produced in generally small amounts in nuclear explosions and nuclear research. They are man-made elements.

Pure substances composed of two or more elements are called compounds. Because they contain two or more elements, compounds, unlike elements, are capable of being decomposed into simpler substances by chemical changes. The ultimate chemical decomposition of compounds produces the elements from which they are made.

The compound carbon monoxide (CO) is composed of carbon and oxygen in the ratio of one atom of carbon to one atom of oxygen. Hydrogen chloride (HCl) contains a ratio of one atom of hydrogen to one atom of chlorine. Compounds may contain more than one atom of the same element. Methane ("natural gas" CH_4) is composed of carbon and hydrogen in a ratio of one carbon atom to four hydrogen atoms. ordinary table sugar (sucrose, $C_{12}H_{22}O_{11}$) contains a ratio of 12 atoms of carbon to 22 atoms of hydrogen to 11 atoms of oxygen. These atoms are held together in the compound by chemical bonds.

There are over three million known compounds, with no end in sight as to the number that can and will be prepared in the future. Each compound is unique and has characteristic physical and chemical properties. Let us consider in some detail two compounds——water and mercuric oxide. Water is a

Words and expressions

element ['ɛləmənt] n. 元素
decompose [ˌdikəm'poz] v. 分解
familiar [fə'mɪljɚ] adj. 熟悉的
oxygen ['ɑksɪdʒən] n. 氧气
aluminum [ə'lumɪnəm] n. 铝
copper ['kɑpə] n. 铜
nitrogen ['naɪtrədʒən] n. 氮
combine [kəm'baɪn] v. 结合

small amounts 少量
nuclear explosion 核爆炸

ultimate ['ʌltəmət] adj. 最终的

monoxide [mə'nɒksaɪd] n. 一氧化物
ratio ['reʃɪo] n. 比率
atom ['ætəm] n. 原子
chloride ['klɔraɪd] n. 氯化物
chlorine ['klɔrin] n. 氯

sucrose ['sjʊkros] n. 蔗糖

chemical bond 化学键

in sight 看得见
unique [ju'nik] adj. 独特的
characteristic [ˌkærəktə'rɪstɪk] adj. 典型的
property n. 性质
in detail 详细地

colorless, odorless, tasteless liquid that can be changed to a solid, ice, at 0℃ and to a gas, steam at 100℃. It is composed of two atoms of hydrogen and one atom of oxygen per molecule, which represents 11.2 percent hydrogen and 88.8 percent oxygen by mass. Water reacts chemically with sodium to produce hydrogen and sodium hydroxide, with lime to produce calcium hydroxide, and with sulfur trioxide to produce sulfuric acid. No other compound has all these exact physical and chemical properties, they are characteristics of water alone.

Mercuric oxide is a dense, orange-red powder composed of a ratio of one atom of mercury to one atom of oxygen. Its composition by mass is 92.6 percent mercury and 7.4 percent oxygen. When it is heated to temperatures greater than 360℃, a colorless gas, oxygen, and a silvery liquid metal, mercury, are produced. Here again are specific physical and chemical properties belonging to mercuric oxide and to no other substance. Thus, a compound may be identified and distinguished from all other compounds by its characteristic properties.

solid ['sɑlɪd] n. 固体
steam [stim] n. 蒸汽
represent [ˌrɛprɪ'zɛnt] v. 代表

sodium ['sodɪəm] n. 钠
lime [laɪm] n. 石灰

dense [dɛns] adj. 密度大的

identify [aɪ'dɛntɪfaɪ] vt. 鉴定
distinguish [dɪ'stɪŋgwɪʃ] v. 辨别

Comprehension

Choose the best answer according to the text.

1. Water is composed of two atoms of hydrogen and one atom of oxygen per molecule, which represents 11.2 percent hydrogen and 88.8 percent oxygen by ( ).
   A. mass　　　B. volume　　　C. mole　　　D. molecule

2. Compounds may contain more than one atom of the same element. The following selections all comply with this sentence except ( ).
   A. CH_4　　　B. HCl　　　C. $C_{12}H_{22}O_{11}$　　　D. CO_2

3. Mercuric oxide is a dense, ( ) powder composed of a ratio of one atom of mercury to one atom of oxygen.
   A. green　　　B. colorless　　　C. orange-red　　　D. black

4. Which statement is not true according to the text? ( )
   A. Compound can be decomposed into simpler substances by chemical changes.
   B. Gold can be found in highly pure forms in nature because it is inert.
   C. Water reacts chemically with lime to produce calcium hydroxide and hydrogen gas.
   D. The formula of mercuric oxide is HgO.

Vocabulary building

Active words

substance
n. [物] 物质；实质

combine
v. 结合

matter
n. 物质；事情

form
v. 形成
n. 形式

nature
n. 自然；性质

produce
v. 生产；产生
n. 农产品，产品

bond
v. 结合
n. 连接，键

react
v. 反应

property
n. 性质；财产

Useful expressions

be composed of 由…组成
be decomposed into 分解成
A reacts with B to produce C and D A 与 B 反应生成 C 和 D

ratio of … to… 比例为……
percent by mass 质量百分比
in sight 在即，在望；看得见

Exercise

Fill in each blank with a given word or expression in their right form.

react   compose   bond   combine   property   solid

1. Tannin is a plant polyphenol. It could _____ protein in solution to form sediment.
2. England, Scotland and Wales _____ the island of Great Britain.
3. An acid can _____ with a base to form a salt.
4. One _____ of red phosphorous is flammable.
5. Matter exists in three states: _____, liquid and gas.
6. These amino acids can react with each other to form a different kind of chemical _____.

Extension

The Elements

Tom Lehrer is an American singer-songwriter, satirist, pianist, and mathematician. *The Elements* is one of his most famous creations, which consists of little more than the elements of the periodic table.

Listen to the song *The Elements*, and try to write down the lyrics.

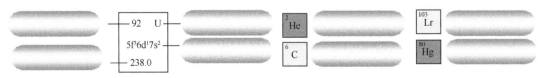

Work in groups

Discuss with your group members about rules of word-formation of following elements.

chlorine   oxygen   sodium   aluminum
carbon   sulfur   fluorine   nitrogen
gold   helium   hydrogen   calcium

Exercise

Match the given words or phrases with the right symbol in the periodic table of the elements.

Atomic weight   Metal   Non-metal   Man-made element
Atomic number   Rare gases   Element symbol   Arrangement of extra
                                                 nuclear electrons

Text B Systematic Nomenclature of Binary Compounds

(1) Binary compounds in which the electropositive element has a fixed oxidation state. The chemical name is composed of the name of the metal, which is written first, followed by the name of the nonmetal, which has been modified to an identifying stem to which is added the suffix-ide. For example, sodium chloride, NaCl is composed of one atom each of sodium and chlorine. The compound name is sodium chloride.

Elements: sodium (metal) + chlorine (nonmetal)
Name of compound: sodium chloride

Stems of some common nonmetals are shown in the following table 1.1.

Table 1.1 Stems of some common nonmetals

Symbol	Element	Stem	Binary name endings
Cl	Chlorine	Chlor	Chloride
H	Hydrogen	Hydr	hydride
N	Nitrogen	Nitr	nitride
O	Oxygen	Ox	oxide

(2) Binary compounds containing metals of variable oxidation numbers and nonmetals. Two systems are commonly used for compounds in this category. The official system, designated by the International Union of Pure and Applied Chemistry (IUPAC), is known as the Stock system. In the Stock System, when a compound contains a metal that can have more than one oxidation number, the oxidation number of the metal in the compound is designated by a Roman numeral in parentheses [e. g. (II)] written immediately after the name of the metal. The nonmetal is treated in the same manner as in the previous case.

Examples:
$FeCl_2$ Iron (II) chloride   (Fe^{2+})
$FeCl_3$ Iron (III) chloride (Fe^{3+})
$CuCl$ Copper (I) chloride (Cu^{+})
$CuCl_2$ Copper (II) chloride (Cu^{2+})

In classical nomenclature, when the metallic

ion has only two oxidation numbers, the name of the metal is modified with the suffixes-ous and-ic to distinguish between the two. The lower oxidation state is given the-ous ending and the higher one the-ic ending.

$FeCl_2$　　ferrous chloride　　　　(Fe^{2+})
$FeCl_3$　　ferric chloride　　　　　(Fe^{3+})
$CuCl$　　　Cuprous chloride　　　(Cu^{+})
$CuCl_2$　　Cupric chloride　　　　(Cu^{2+})

(3) Binary compounds contain two nonmetals. The most electropositive element is named first. Between two nonmetals, the element that occurs earlier in the following sequence is written and named first in the formula: B, Si, C, P, N, S, I, Br, Cl, O, F. To each element is attached a Latin or Greek prefix indicating the number of atoms of that element in the molecule. The second element still retains the modified binary ending. The prefix mono is generally omitted except when needed to distinguish between two or more compounds. Common prefixes and their numerical equivalences are the following.

mono = 1　di = 2　tri = 3　tetra = 4　penta = 5
hexa = 6　hepta = 7　octa = 8　nona = 9　deca = 10

Examples:

CO　　carbon monoxide
CO_2　　carbon dioxide
PCl_3　　phosphorus trichloride
PCl_5　　phosphorus pentachloride

(4) Exceptions using-ide endings. Three notable exceptions using the-ide ending are hydroxides (OH^-), cyanides (CN^-), and ammonium (NH_4^+) compounds.

NH_4I　　　　Ammonium iodide
$Ca(OH)_2$　　Calcium hydroxide
KCN　　　　Potassium cyanide

(5) Acids derive from binary compounds. Certain binary hydrogen compounds, when dissolved in water, form solutions that have acid properties. Because of this property, these compounds are given acid names in addition to their regular-ide names. However, not all binary hydrogen compounds are acids. To express the formula of a binary acid, it is customary to write the symbol of hydrogen first, followed by the symbol of the second element (e. g. HCl, HBr, H_2S).

To name a binary acid, place the prefix hydro- in front of, and the letter-ic after the stem of the

suffix ['sʌfɪks] n. 后缀

formula ['fɔrmjələ] n. 分子式
attach [ə'tætʃ] v. 附加
prefix ['prifɪks] n. 前缀

retain [rɪ'ten] v. 保留
omit [ə'mɪt] v. 省略

equivalence [ɪ'kwɪvələns] n. 相等

derive [dɪ'raɪv] v. 源于
dissolve [dɪ'zɑlv] v. 溶解
solution [sə'luʃən] n. 溶液

customary ['kʌstə'mɛri] adj. 习惯的

nonmetal. Then add the word "acid".
  Examples: HCl Hydrochloric acid
           H_2S Hydrosulfuric acid

Comprehension

Choose the best answer according to the text.

1. According to the text, which compound owns multiple? names? (　　)
   A. CO_2　　　　　B. HCl　　　　　C. NaCl　　　　　D. KCN
2. Which one belongs to binary compounds only containing two nonmetals? (　　)
   A. CO_2　　　　　B. Ca(OH)$_2$　　C. NaCl　　　　　D. KCN
3. Two systems are commonly used for binary compounds containing metals of variable oxidation numbers and nonmetals. They are (　　).
   A. official system and Stock system
   B. official system and classical nomenclature system
   C. official system and IUPAC
   D. IUPAC and classical nomenclature system
4. Which statement is not true according to the text? (　　)
   A. H_2S dissolved in water will form solutions that have acid properties.
   B. HCl is named as hydrochloric acid or hydrogen chloride.
   C. CO is named as carbon oxide because the prefix mono is generally omitted.
   D. Ferric chloride and iron (III) chloride are the same chemical.

Match the formulae with their equivalents

1. SO_3 A. sodium hydroxide
2. CCl_4 B. carbon tetrachloride
3. NH_4Cl C. hydrogen sulfide
4. NaOH   D. ammonium chloride
5. H_2S E. sulfur trioxide
6. CO   F. carbon monoxide

Write the systematic nomenclature of the following chemicals.

H_2 _____   KOH _____
Ca _____   HCl _____
CO _____ SO_2 _____
NaCl _____ PCl_3 _____
Fe^{3+} _____ PCl_5 _____
Fe^{2+} _____ $Ca(OH)_2$ _____

Vocabulary building

Active words

variable   n. [数] 变量
adj. 可变的

formula
n. 公式；分子式；配方

attach
v. 附属，附加

retain
v. 保持

omit
vt. 省略；遗漏；未（做）

exception
n. 例外

express
vt. 表达
n. 快车，快递

dissolve
vt. 溶解；解散，解除，消散

Useful expressions

consist of　由…组成；由…构成；包括
be known as　被称为；被认为是
derive from　来源于；衍生于
in addition to　除了

Exercise

Fill in each blank with a given word or expression in their right form.

in addition to　　consist of　　variable　　be known as　　dissolve　　express

1. Solubility is the degree to which a substance dissolves in a solvent to make a solution (usually _____ as grams of solute per liter of solvent).

2. Warm the sugar slightly first to make it _____ quicker.

3. He is of a _____ mood; he never finishes what he starts.

4. Mercury is the metal which is liquid at room temperature and is better _____ quicksilver.

5. _____ Sinopec, 37 other companies have also been punished for water pollution.

6. The atmosphere _____ more than 70% of nitrogen.

Work in groups

Discuss with your group members about nomenclature of following organic substances and find out the common suffixes and prefixes.

Formula	System nomenclature	Formula	System nomenclature
CH_4	methane	CH_3-	methyl
C_2H_6	ethane	CH_3CH_2-	ethyl
C_3H_8	propane	$CH_3CH_2CH_2-$	propyl
C_4H_{10}	butane	CH_3OH	methanol
C_5H_{12}	pentane	$CH_3(CH_2)_3CH_2-$	pentyl
C_6H_{14}	hexane	$CH_2=CH_2$	ethylene
C_7H_{16}	heptane	$CH_3CH=CH_2$	propene
C_8H_{18}	octane	$CH_3CH_2CH_2CH=CH_2$	1-pentene
C_9H_{20}	nonane	$HC\equiv CH$	ethyne
$C_{10}H_{22}$	decane	$CH_3C\equiv CH$	propyne

Write the nomenclature of the following chemicals.

CH_4 _____

$CH_3CH_2CH_2CH_3$ _____

C_2H_5OH _____

C_6H_6 _____

Text C Properties of Aqueous Ammonia

Words and expressions

Ammonia is an important raw material for chemical industry and has a wide use in various industries. Therefore, the properties of ammonia are of importance.

aqueous ['ekwɪəs] adj. 水的
ammonia [ə'monɪə] n. [无化] 氨

Section 1 Physical and Chemical Properties

Appearance and Characters: Colorless liquid.
Melting Point (℃): -77.7
Relative Density (1 for water): 0.82/-79℃
Boiling Point (℃): -33.5
Relative Vapor Density (1 for air): 0.6
Saturated Vapor Pressure (kPa): 506.42/4.7℃
Combustion heat (kJ/mol): No data available.
Critical temperature (℃): 132.5
Critical pressure (MPa): 11.40
Logarithmic Value of Octanol/Water Distribution Coefficient: No data available.
Flash Point (℃): No data available.
Ignition Point (℃): 651
Upper Limit of Explosion, % (V/V): 27.4
Lower Limit of Explosion, % (V/V): 15.7

Solubility: The product dissolves sodium compound, potassium compound, sulfur compound and phosphorus compound, inorganic chloride, bromide, sulfonated bodies, cyanide, nitrate, nitrite, organic amine compound, phenol, alcohol and aldehyde, etc. Main Applications: The product is mainly used to produce fertilizer, and can be used as fertilizer directly. In industry it is mainly used to produce dynamite, various chemical fibers and plastics. They can be used as refrigerant as well. It is widely used in wood paper pulp production, metallurgy, oil refining, rubber, leather manufacture and medicine, etc.

vapor ['veipə] n. 蒸气
saturate ['sætʃərit] vt. 使饱和
combustion [kəm'bʌstʃən] n. 燃烧
critical ['krɪtɪkl] adj. 临界的
logarithmic [lɔgə'rɪðmɪk] adj. 对数的
distribution coefficient [物化] 分配系数
ignition [ɪg'nɪʃən] n. 燃烧
explosion [ɪk'sploʒən] n. 爆炸
solubility [ˌsɑljə'bɪləti] n. 溶解度
potassium [pə'tæsɪəm] n. 钾
phosphorus ['fɑsfərəs] n. 磷
cyanide ['saɪə,naɪd] n. 氰化物
amine [ə'min] n. [有化] 胺
phenol ['finɔl] n. 苯酚
aldehyde ['ældə,haɪd] n. 醛
fertilizer ['fɜtəlaɪzɚ] n. 肥料
dynamite ['daɪnə'maɪt] n. 炸药
fiber ['faɪbɚ] n. 纤维
refrigerant [ri'frɪdʒərənt] n. 制冷剂
pulp [pʌlp] n. 纸浆
metallurgy ['mɛtələdʒi] n. 冶金
stability [stə'bɪləti] n. 稳定性

Section 2 Stability and Reactivity

Stability: Stable at ambient temperature.

Product Development | 11

Incompatibility: Halogen, acyl chloride, acid, chloroform and strong oxidants.

Contact Avoiding Conditions: No data available.

Danger of Polymerization: Impossible to occur.

Decomposition Products: Hydrogen and nitrogen.

Section 3　Toxicology Data

Toxicity: The product is slightly toxic. It has an irritant and corrosive effect on the upper respiratory tract. The product with high concentration may increase the excitability of central nervous system, causing spasms. Ammonia and aqua ammonia may cause eye corneal edema, even corneal perforation. Aqua ammonia water may cause skin ambustion as well.

ambient ['æmbɪənt] adj. 周围的
incompatibility [ˌɪnkəmˌpætə'bɪləti] n. 不相容
halogen ['hælədʒən] n. 卤素
acyl chloride 酰基氯
toxicology [ˌtɑksɪ'kɑlədʒi] n. 毒物学
irritant n. [医] 刺激物, [医] 刺激剂
corrosive [kə'rosɪv] adj. 腐蚀的
respiratory tract 呼吸道
spasms n. 肌痉挛
eye corneal edema 眼角膜水肿
corneal perforation 角膜穿孔
ambustion [æm'bʌstʒən] n. 灼伤

Comprehension

Fill in the blanks with the given words.

The passage is concerned with the properties of aqueous ammonia, which have great bearing on the safe handling procedures for the chemical. Let's consider some properties in detail.

Boiling point is an indicator of how readily the chemical becomes a _____ (gas/liquid/solid). The _____ (lower/higher) the boiling point, the more readily it vaporizes. The boiling point of ammonia is $-33.5℃$, so it should be in _____ (gas/liquid/solid) state at room temperature.

The lower explosive limit (LEL) is the lowest concentration of solvent in air that will ignite. The upper explosive limit (UEL) is the highest concentration of solvent in air that will ignite. As a rule of thumb, the greater the range between the LEL and UEL, the greater the fire hazard. For example:

Ether　　　　　　　　LEL=1.9%. UEL=36%
Aqueous ammonia　　LEL=15.7%. UEL=27.4%

Based on these values only, ether presents a _____ (greater/less) fire hazard than aqueous ammonia. However, to determine the fire hazard accurately, flash point and vapor pressure would also need to be considered.

A vapor which is heavier than air (vapor density greater than 1) will tend to collect in pools and spread near ground level. A vapor which is lighter than air will tend to rise. Aqueous ammonia belong to the _____ (former/latter).

Vocabulary building

Active words

available
adj. 可得的；可用的

concentration
n. 浓度；集中；专心

critical
adj. 临界的；关键的；批评的

manufacture
n. 制造

v. 制造

various
adj. 各种各样的

application
n. 应用；申请

body
n. 身体；主体

Useful expressions
has an effect on 对……有影响

as well 也

Exercise

Match the parameters with their equivalents.

1. melting point   A. 临界压力
2. flash point   B. 着火点
3. relative density   C. 闪点
4. ignition point   D. 相对密度
5. critical pressure   E. 蒸气压
6. vapor pressure   F. 熔点

Translate the specification sheet into Chinese.

Properties of HAP

[usage]: mainly used in resin industry, used as dispersant, such as EPS, SAN, PS. The anti-adhesive separating agent, also applied in ceramic, biological material, fluorescent material etc.

[scientific name]: HAP

[molecular formula]: $Ca_2(PO_4)_6(OH)_2$

[properties]: White amorphous powder

[relative density]: 3.18

[boiling point]: 1670℃

[characteristics]: even grain distribution, stable production, easy control. No change the process and equipment, this product can be applied in suspension polymerization. Quality product in low price.

[technical index]:

item	index	item	index
P_2O_5	38.0%～41.7%	pH	6.0-9.0
CaO	50.0%～53.5%	Insoluble matter in HCl	≤0.3%
moisture	≤5.0%		

[package & storage]: 10kg/20kg in woven bag lined with double layers polythene bag. Keep in cool airiness dry place and avoid sunlight, rain and breakage.

Extension

How to get the useful information of chemicals?

Lange's Handbook of Chemistry

Lange's Handbook of Chemistry is a chemical data handbook with complete and accurate data, especially convenient for chemical and related scientific workers.

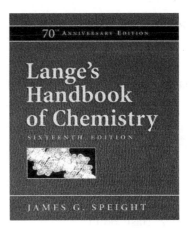

The book was compiled by N. A. Lange (1st-10th edition), followed by J. A. Dean (11th-15th edition). The latest 16th edition was edited by J. G. Speight.

The book (16th edition) is divided into eleven sections:

Section 1. Organic Compounds

Section 2. General Information, Conversion Tables, and Mathematics

Section 3. Inorganic Compounds

Section 4. Properties of Atoms, Radicals, and Bonds

Section 5. Physical Properties

Section 6. Thermodynamic Properties

Section 7. Spectroscopy

Section 8. Electrolytes, Electromotive Force, and Chemical Equilibrium

Section 9. Physicochemical Relationships

Section 10. Polymers, Rubbers, Fats, Oils, and Waxes

Section 11. Practical Laboratory Information

The following table is extracted from Lange's Handbook of Chemistry. According to the data in Table 1.2, answer questions:

1. How many drying agents are listed and what are they?
2. Which drying agents can be used for drying methane?
3. As a drying agent, K_2CO_3 is useful for most materials except acids, why?
4. which drying agent can be used only once?

Table 1.2 Drying agents

Drying agent	Most useful for	Residual water, mg H_2O per liter of dry air(25℃)	Grams water removed per gram of desiccant	Regeneration/℃
Al_2O_3	Hydrocarbons	0.002-0.005	0.2	175(24h)
CaO	Ethers, esters, alcohols, amines	0.01-0.003	0.31	Difficult, 1000
$CaSO_4$	Most organic substances	0.005-0.07	0.07	225
K_2CO_3	Most materials except acids and phenols		0.16	158
KOH	Amines	0.01-0.9		Impossible

> **Reading material**

The Haber Process for Ammonia Synthesis

All methods for making ammonia are basically fine-tuned versions of the process developed by Haber, Nemst and Bosch in Germany just before the First World War.

In principle the reaction between hydrogen and nitrogen is easy; it is exothermic and the equilibrium lies to the right at low temperatures. Unfortunately, nature has bestowed dinitrogen with an inconveniently strong triple bond, enabling the molecule to thumb its nose at thermodynamics. In scientific terms the molecule is kinetically inert, and rather severe reaction conditions are necessary to get reactions to proceed at a respectable rate. A major source of "fixed" (meaning, paradoxically, "usefully reactive") nitrogen in nature is lightning, where the intense heat is sufficient to create nitrogen oxides from nitrogen and oxygen.

To get a respectable yield of ammonia in a chemical plant we need to use a catalyst. What Haber discovered, and it won him a Nobel Prize, was that some iron compounds were acceptable catalysts. Even with such catalysts extreme pressures (up to 600 atmospheres in early processes) and temperatures (perhaps 400℃) are necessary.

Pressure drives the equilibrium forward, as four molecules of gas are being transformed into two. Higher temperatures, however, drive the equilibrium the wrong way, though they do make the reaction faster; chosen conditions must be a compromise that gives an acceptable conversion at a reasonable speed. The precise choice will depend on other economic factors and the details of the catalyst. Modern plants have tended to operate at low pressures and higher temperatures (recycling unconverted material) than the nearer-ideal early plants, since the capital and energy costs have become more significant.

Biological fixation also uses a catalyst which contains molybdenum (or vanadium) and iron embedded in a very large protein. The detailed structure of which eluded chemists until late 1992. How it works is still not understood in detail.

The common features of all the different varieties of ammonia plant are that the synthesis gas mixture is heated, compressed and passed into a reactor containing a catalyst. The essential equation for the reaction is simple:

$$N_2 + H_2 \longrightarrow NH_3$$

Early plants plumped for very high pressure (to get the yield up in a one-pass reactor), but many of the most modem plants have accepted much lower one-pass yields at lower pressures and have also opted for lower temperatures to conserve energy. LCA (leading concept ammonia) process, for example, works at very low pressures (for an ammonia plant) and recycles unreacted gases, paying careful attention to the energy flow in the process so that energy produced at one stage is not wasted by dissolution to atmosphere, but is used for other stages requiring energy input. The LCA process also uses a highly active long-life catalyst which is protected from poisons by very careful purification of the input gases.

In order to ensure the maximum yield in the reactor the synthesis gas is usually cooled as it reaches equilibrium. The effect of this is to freeze the reaction as near to equilibrium as possible. Since the reaction is exothermic (and the equilibrium is less favorable for ammonia synthesis at higher temperatures), the heat must be carefully controlled in this way to achieve good yields.

The output from the Haber stage will consist of a mixture of ammonia and synthesis gas; so the next stage needs to be the separation of the two so that the synthesis gas can be recycled. This is normally accomplished by condensing the ammonia (which is a good deal less volatile than the other components; ammonia boils at about $-40℃$).

The major use of ammonia is not for the production of nitrogen-containing chemicals for further industry use, but for fertilizers such as urea or ammonium nitrates and phosphates. Fertilizers consume 80% of all the ammonia produced. In USA in 1991, for example, the following ammonia-derived products were consumed, mostly for fertilizers: urea, ammonium sulfate, ammonium nitrate and diammonia hydrogen phosphate.

Chemical uses of ammonia are varied. The Solvay process for the manufacture of soda ash uses ammonia, though it does not appear in the final product since it is recycled. A wide variety of processes are taken in ammonia directly, including the production of cyanides and aromatic nitrogen-containing compounds such as pyridine. The nitrogen in many polymers (such as nylon or acrylics) can be traced back to ammonia, often via nitriles or hydrogen cyanide. Most other processes use nitric acid or salts derived from it as their source of nitrogen. Ammonium nitrate, used as a nitrogen-rich fertilizer, also finds a major use as a bulk explosive.

Comprehension

Answer the following questions according to the passage.

1. Why severe temperature conditions are adopted in plants since the reaction equilibrium lies to the right at low temperatures?
2. What are the common catalysts to produce ammonia?
3. What are the common features of all the different varieties of ammonia plant?
4. How to control the energy flow to ensure the maximum yield in the reactor?
5. What are the main usages of ammonia?

Supplementary knowledge

Structure of Patent Documents

A patent document consists of four parts: title page, description, claims and drawings. These elements are obligatory for all patent documents, although the drawings may be omitted if the subject matter of the invention does not require them.

Title page (see Fig. 1.1)

The title page of a patent document is therefore of central importance. It contains

(1) the title of the invention,

(2) the bibliographic data,

(3) an abstract which provides initial technical information, and,

(4) if applicable, a basic drawing which illustrates the invention.

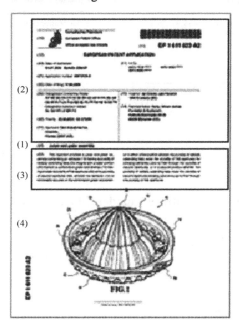

Fig. 1.1　The title page

Bibliographic data (see Fig. 1.2)

The most important bibliographic data is

(10) or (11) the publication number,

(21) the application number,

(22) the date of filing at the patent office,

(71) the applicant,

(30) data relating to priority,

(72) the inventor.

Description

The description describes all aspects of the subject matter of the invention. Patent law demands that the invention must be described in sufficient detail for it to be carried out by a person skilled in the art.

Product Development | 17

Fig. 1.2 The bibliographic data

Claims

The claims are the most important part of the document: the "legal core", as it were. They define the features of the invention to be protected by the patent.

Drawings

Almost all patent and utility model documents contain one or more drawings. These are usually black and white line drawings and are often very helpful for understanding the idea behind the invention.

Adapted from http: //www. epo. org/

Exercise

Read the patent file quickly and answer the questions.

1. Who is the inventor?
2. Who is the applicant?
3. What is the title of the invention?
4. What is publication number?
5. What is the content of the invention?

(12) INTERNATIONAL APPLICATION PUBLISHED UNDER THE PATENT COOPERATION TREATY (PCT)

(19) World Intellectual Property
Organization
International Bureau

(43) International Publication Date
31 December 2014 (31.12.2014)

(10) International Publication Number
WO 2014/209643 A1

(51) International Patent Classification:
B01J 3/00 (2006.01) *B01J 12/00* (2006.01)
B01J 8/02 (2006.01)

(21) International Application Number:
PCT/US2014/042445

(22) International Filing Date:
15 June 2014 (15.06.2014)

(25) Filing Language: English

(26) Publication Language: English

(30) Priority Data:
61/839,001 25 June 2013 (25.06.2013) US
61/782,719 1 September 2013 (01.09.2013) US
61/877,994 15 September 2013 (15.09.2013) US
61/890,690 14 October 2013 (14.10.2013) US
61/916,915 17 December 2013 (17.12.2013) US
61/919,786 22 December 2013 (22.12.2013) US

(72) Inventor; and
(71) Applicant : GOLDSTEIN, Leonid [IL/US]; 12501 Tech Ridge Blvd., Apt. 1535, Austin, Texas 78753 (US).

(81) Designated States *(unless otherwise indicated, for every kind of national protection available)*: AE, AG, AL, AM, AO, AT, AU, AZ, BA, BB, BG, BH, BN, BR, BW, BY, BZ, CA, CH, CL, CN, CO, CR, CU, CZ, DE, DK, DM, DO, DZ, EC, EE, EG, ES, FI, GB, GD, GE, GH, GM, GT, HN, HR, HU, ID, IL, IN, IR, IS, JP, KE, KG, KN, KP, KR, KZ, LA, LC, LK, LR, LS, LT, LU, LY, MA, MD, ME, MG, MK, MN, MW, MX, MY, MZ, NA, NG, NI, NO, NZ, OM, PA, PE, PG, PH, PL, PT, QA, RO, RS, RU, RW, SA, SC, SD, SE, SG, SK, SL, SM, ST, SV, SY, TH, TJ, TM, TN, TR, TT, TZ, UA, UG, US, UZ, VC, VN, ZA, ZM, ZW.

(84) Designated States *(unless otherwise indicated, for every kind of regional protection available)*: ARIPO (BW, GH, GM, KE, LR, LS, MW, MZ, NA, RW, SD, SL, SZ, TZ, UG, ZM, ZW), Eurasian (AM, AZ, BY, KG, KZ, RU, TJ, TM), European (AL, AT, BE, BG, CH, CY, CZ, DE, DK, EE, ES, FI, FR, GB, GR, HR, HU, IE, IS, IT, LT, LU, LV, MC, MK, MT, NL, NO, PL, PT, RO, RS, SE, SI, SK, SM, TR), OAPI (BF, BJ, CF, CG, CI, CM, GA, GN, GQ, GW, KM, ML, MR, NE, SN, TD, TG).

Declarations under Rule 4.17:
— *as to the identity of the inventor (Rule 4.17(i))*
— *as to applicant's entitlement to apply for and be granted a patent (Rule 4.17(ii))*
— *as to the applicant's entitlement to claim the priority of the earlier application (Rule 4.17(iii))*
— *of inventorship (Rule 4.17(iv))*

[Continued on next page]

(54) Title: SYSTEM AND METHOD FOR PERFORMING ENDOTHERMIC REACTIONS

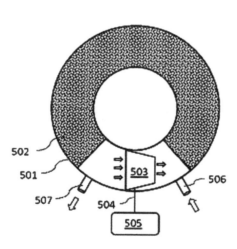

(57) Abstract: A system and a method for producing synthesis gas (syngas) and performing other catalyst based endothermic chemical reactions in gases in industrial settings, supplying external energy in the form of mechanical energy. The mechanical energy becomes heat through one of the following non-alternative processes in the gas: compression, pressure drop, velocity drop, friction with solid surface and turbulence. Multiple examples and applications for producing syngas, methanol, synthetic gasoline, hydrogen and ammonia are disclosed.

Mass Production

Objectives:

After finishing this module, you are able to:
- Choose types of chemical reactors according to the information
- Understand the principles of unit operation
- Use equipment manuals properly

Warming-up

Development of Chemical Engineering

Chemical engineers

It was Edward Charles Howard who was rewarded as "the first chemical engineer of any eminence". Chemical engineers use chemistry and engineering to turn raw materials into usable products, such as medicine, petrochemicals and plastics on a large-scale, industrial setting.

Sulfuric acid

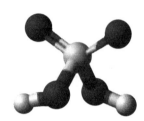

One of the first chemicals to be produced in large amounts through industrial process, was sulfuric acid which is also a central substance in the chemical industry. Principal uses include mineral processing, fertilizer manufacturing, oil refining, wastewater processing, and chemical synthesis.

Chemical enterprise

The largest corporate producers worldwide, each with plants in numerous countries, include BASF, Bayer, Dow Chemical, DuPont, Eastman Chemical Company. Thereare thousands of smaller firms. The United States alone produced $689 billion, 18.6 percent of the total world chemical output in 2008.

Unit operation

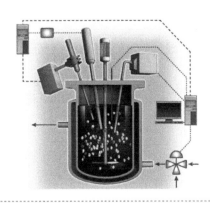

A unit operation is a physical step in an individual chemical engineering process. Unit operations (such as crystallization, filtration, drying and evaporation) are used to prepare reactants, to purify and separate its products, to recycle unspent reactants, and to control energy transfer in reactors.

Text A Types of Chemical Reactors

The number of types of reactors is very large in the chemical industry. Even for the same operation, such as nitration of toluene, different types are used: the batch reactor, the continuous stirred tank, and a cascade of stirred tanks.

Batch (see Fig. 2.1) and Semibatch Reactors

Fig. 2.1 Batch reactor

Batch reactors are generally used for liquid phase reactions. When a solid catalyst has to be kept in suspension or when there are two liquid phases, as in the nitration of aromatics, for instance, an agitator is required. Agitation is also necessary to equalize the temperature in the reactor and to keep it at the desired level by efficient heat exchange through a jacket or an internal coil.

In pure batch operation the reactants are completely fed into the reactor at the beginning. For better control of temperature this type of operation may not be advisable and the reactant(s) may have to be added progressively to the contents of the vessel. The reactor is then said to operate in the semibatch mode.

Batch and semibatch reactors are most often used for low production capacities. They are generally encountered in the area of specialty chemicals and polymers and in pharmaceuticals, in particular, in plants with a wide variety of products.

The Plug Flow Reactor (see Fig. 2.2)

Plug flow is a simplified and idealized picture of the motion of a fluid, whereby all the fluid elements move with a uniform velocity along parallel streamlines. This perfectly ordered flow is the only transport mechanism accounted for in the plug flow reactor model. Because of the uniformity of conditions in a cross section, the steady-state continuity equation is a very simple ordinary differential equation.

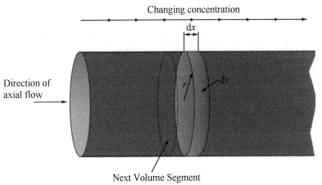

Fig. 2.2 The plug flow reactor

cross section 横剖面
continuity ['kɒntə'nuəti] n. 连续性
differential ['dɪfə'renʃəl] adj. 微分的
equation [ɪ'kweʒn] n. 方程式

The Perfectly Mixed Flow Reactor

This reactor type is the opposite extreme of the plug flow reactor. In the perfectly mixed flow reactor, the conversion takes place at a unique concentration (and temperature) level which, of course, is also the concentration of the effluent.

The stirred flow reactor see Fig. 2.3 is frequently chosen when temperature control is a critical aspect, as in the nitration of aromatic hydrocarbons or glycerine.

extreme [ɪk'stri:m] n. 极端

effluent ['efluənt] n. 水流
aspect ['æspekt] n. 方面
glycerine ['glɪsəri:n] n. 甘油，丙三醇

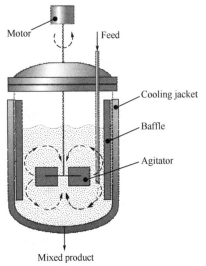

Fig. 2.3 The stirred flow reactor

Several alternate names have been used for what is called here the perfectly mixed flow reactor. One of the earliest was "continuous stirred tank reactor," or CSTR, which some have modified to "continuous flow stirred tank reactor,"

alternate ['ɔltə-nət] adj. 替代的

continuous stirred tank reactor 连续搅拌釜式反应器

or CFSTR.

Fixed Bed Catalytic Reactors

The discovery of solid catalysts and their application to chemical processes in the early years of the 20th century has led to a breakthrough of the chemical industry. Since then, this industry has diversified and grown in a spectacular way through the development of new or the rejuvenation of established processes, mostly based on the use of solid catalysts.

fixed bed catalytic reactors 固定床催化反应器

breakthrough ['brek'θru] n. 突破
diversify [daɪ'vɜsɪfaɪ] v. 多样化
rejuvenation [rɪ,dʒuvə'heʃən] n. 复苏

Adapted from Chemical Reactor Analysis and Design (3rd Edition)
by Gilbert F. Froment etc.

Comprehension

Choose the best answer according to the text.

1. Which one is not the operating condition of the reactor used for liquid phase reactions? ( )
   A. Compositions   B. Flow rates   C. Pressure   D. Temperature
2. Which part of batch reactors serves as a heat exchange? ( )
   A. Jacket   B. Agitator   C. Coil   D. Both A and C
3. Under constant operating conditions, the concentration of reactants remain the same only in ( ).
   A. batch reactors   B. the Plug Flow Reactors
   C. the Perfectly Mixed Flow Reactors   D. fixed Bed Catalytic Reactors
4. Which statement is not true according to the text? ( )
   A. Continuous distillation is an ongoing distillation in which a liquid mixture is continuously (without interruption) fed into the process and separated fractions are removed continuously as output streams as time passes during the operation.
   B. Batch reactors is good for liquid reactions.
   C. Batch and semibatch reactors are most often used for large production capacities.
   D. In chemical engineering, chemical reactors are containers designed to contain chemical reactions.

Vocabulary building

Active words

reaction
n. 反应；(化学) 反应

jacket
n. 夹克；文件套；[化学] 夹套

plant
n. 植物，庄稼；设备；工厂

equation
n. [数] 方程式，等式；[化学] 反应式

concentration
n. 专心；浓度；集中

whereby
adv. 通过……；借以；与……一致

uniform
n. 制服
adj. 一样的

alternate
adj. 交替的；轮流的

Exercise

Fill in each blank with a given word or expression in their right form.

batch   catalyst   reactant   plant   nitration   concentration

1. There was a terrible explosion at the chemical _____.
2. The _____ process has been widely applied in many industries such as the fine chemical industry and the catalyst preparation, etc.
3. What is the _____ of salt in sea water?
4. The reaction takes place only in the presence of a _____.
5. The influence of reacting temperature and ratio of _____ were studied.
6. Some people proposed polymerizing the glycerine prior to _____.

Extension

Fluidized Bed Reactor

The solid substrate (the catalytic material upon which chemical species react) material in the fluidized bed reactor is typically supported by a porous plate, known as a distributor. The fluid is then forced through the distributor up through the solid material. At lower fluid velocities, the solids remain in place as the fluid passes through the voids in the material. This is known as a packed bed reactor. As the fluid velocity is increased, the reactor will reach a stage where the force of the fluid on the solids is enough to balance the weight of the solid material. This stage is known as incipient fluidization and occurs at this minimum fluidization velocity. Once this minimum velocity is surpassed, the contents of the reactor bed begin to expand and swirl around much like an agitated tank or boiling pot of water. The reactor is now a fluidized bed. Depending on the operating conditions and properties of solid phase various flow regimes can be observed in this reactor.

Work in groups

Discuss how to select reactors according to the state of reactants, and complete the following table.

Reactor types	the state of reactants
Tank reactor	
Tube reactor	
Fixed Bed Catalytic Reactors	
Fluidized bed reactor	

Text B Basic Distillation Equipment and Operation

Words and expressions

Distillation is defined as a process in which a liquid or vapor mixture of two or more substances is separated into its component fractions of desired purity, by the application and removal of heat.

One way of classifying distillation column type is to look at how they are operated. Thus we have batch and continuous columns.

Distillation columns are made up of several components, each of which is used either to transfer heat energy or enhance material transfer. A typical distillation contains several major components:

➡ a vertical shell where the separation of liquid components is carried out

➡ column internals such as trays/plates and/or packings which are used to enhance component separations

➡ a reboiler to provide the necessary vaporization for the distillation process

➡ a condenser to cool and condense the vapor leaving the top of the column

➡ a reflux drum to hold the condensed vapor from the top of the column so that liquid (reflux) can be recycled back to the column

A schematic of a typical distillation unit with a single feed and two product streams is shown in Fig. 2.4:

distillation [ˌdɪstl'eʃən] n. 蒸馏
vapor ['vepɚ] n. 蒸汽

column ['kɑləm] n. 圆柱

enhance [ɪn'hæns] n. 增强

vertical ['vɝtɪkl] adj. 垂直的
shell [ʃɛl] n. 壳体
internals [ɪn'tɝːnlz] n. 内部构件
tray [tre] n. 塔盘
plate n. 塔板
packing n. 填料
reboiler n. 再沸器
vaporization [ˌvepərɪ'zeʃən] n. 汽化
condense [kən'dɛns] vt. 使冷凝
reflux ['riˌflʌks] n. 回流
reflux drum 回流罐
schematic [ski'mætɪk] n. 示意图

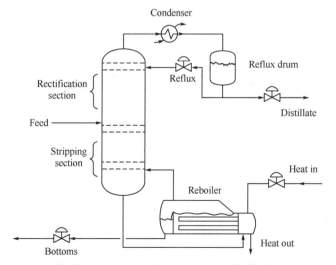

rectification section 精馏段
stripping section 提馏段

Fig. 2.4 A schematic of a typical distillation

The liquid mixture that is to be processed is known as the feed and this is introduced usually somewhere near the middle of the column to a tray known as the feed tray. The feed tray divides the column into a top (enriching or rectification) section and a bottom (stripping) section. The feed flows down the column where it is collected at the bottom in the reboiler (See Fig. 2.5).

feed tray [化工] 进料塔盘
enrich [ɪnˈrɪtʃ] vt. 浓缩
rectification [ˌrɛktəfəˈkeʃən] [化工] 精馏

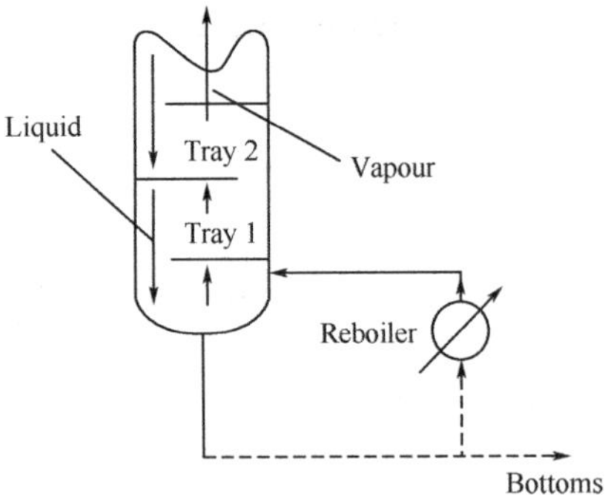

Fig. 2.5　The Reboiler

Heat is supplied to the reboiler to generate vapor. The source of heat input can be any suitable fluid, although in most chemical plants this is normally steam. In refineries, the heating source may be the output streams of other columns. The vapor raised in the reboiler is reintroduced into the unit at the bottom of the column. The liquid removed from the reboiler is known as the bottoms product or simply, bottoms.

input [ˈɪnˈpʊt] n. 给料

output [ˈaʊtpʊt] n. 出料

The vapor moves up the column, and as it exits the top of the unit, it is cooled by a condenser (See Fig. 2.6). The condensed liquid is stored in a holding vessel known as the reflux drum. Some of this liquid is recycled back to the top of the column and this is called the reflux. The condensed liquid that is removed from the system is known as the distillate or top product.

condenser n. 冷凝器

distillate [ˈdɪstɪleɪt] n. 馏出液

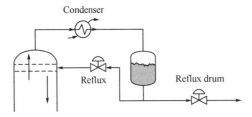

Fig. 2.6　The condenser

Thus, there are internal flows of vapor and liquid within the column as well as external flows of feeds and product streams, into and out of the column.

Adapted from Distillation by M. T. Tham

Comprehension

Answer the questions in English.

1. Could you give a brief explanation of distillation?
2. Could you list some usages of distillation?
3. In batch distillation equipment, what are necessary components?
4. Could you describe the working procedures of continuous distillation?

Decide whether the following statements are "true" or "false" according to the passage.

(   ) 1. Distillation is used to separate a liquid or vapor mixture into its components by the use of heat.

(   ) 2. After distillation, there will be no volatile materials contained in the original mixture.

(   ) 3. Heat is supplied to the reboiler to generate vapor.

(   ) 4. In the column, vapor flows up contacting liquid at each tray.

(   ) 5. The hottest tray is on the top of the column.

(   ) 6. The feed tray and the trays blow are called stripping section.

Vocabulary building

Active words

fraction
n. [数] 分数；[化工] 馏分

feed
vt. 喂养
n. 饲料；[化工] 进料；原料

condense
v. 压缩，浓缩；（使）冷凝，凝结

process
n. 过程；工序
vt. 加工，处理

remove
vt. 去除

vapor
n. 蒸汽
vi. （使）蒸发

steam
n. 蒸汽

generate
vt. 产生

internal
adj. 内部的

Useful expressions

be defined as   被定义为
be made up of   由……组成

be known as   被认为是

Exercise

Fill in each blank with a given word or expression in their right form.

plate   vaporization   distillation   reflux   fraction   distillate

1. By using different collection vessels, the original mixture can be separated into _____.

2. In distillation operation, more theoretical _____ lead to better separations.

3. _____ is used to separate crude oil into more fractions for specific uses such as transport, power generation and heating.

4. Part of the overhead stream is withdrawn as _____, or overhead product.

5. _____ is a phase transition from the liquid phase to vapor.

6. _____ refers to the portion of the condensed overhead liquid product from a distillation or fractionation tower that is returned to the upper part of the tower.

Extension

The Internal Structure of a Packing Tower

1   schematic drawing of packed column (tower) assembly 填料塔装配简图
2   earth lug 接地板
3   pipe opening and support plate 引出孔和支承板
4   liquid outlet 液体出口
5   ellipsoidal head 椭圆形封头
6   manhole 人孔
7   packing 填料
8   shell 壳体
9   handhole 手孔
10   liquid inlet 液体入口
11   top davit 塔顶吊柱
12   gas outlet 气体出口
13   lifting lug 吊耳
14   liquid distributor 液体分布器
15   hole-down grid 格栅式填料压板
16   packing support plate 填料支承板
17   liquid collector 液体收集器
18   by-product 副产品
19   feed for distillation 蒸馏进料
20   liquid redistributors 液体再分布器
21   combined packing support plate and liquid redistributors 组合的填料支承板与液体再分布器
22   gas inlet 气体入口
23   internal flange and piping 内部法兰和接管
24   vent hole (裙座) 排气管 (口)
25   skirt 裙座
26   skirt access opening (access hole) 裙座检查孔

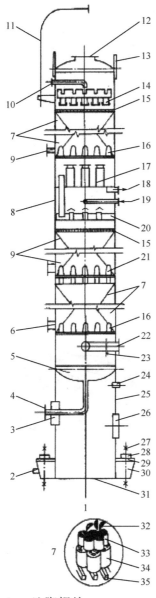
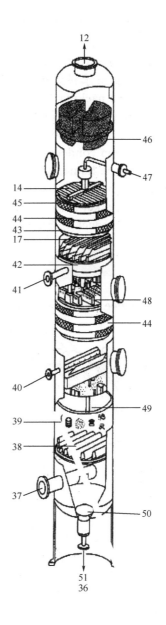

27　anchor 地脚螺栓
28　washer 垫板
29　compression ring 压环，环形盖板
30　gusset plate 筋板
31　base ring 基础环，底板
32　small packing 小型填料
33　medium packing 中型填料
34　large packing 大型填料
35　support bar 支承杆
36　diagrammatic sketch of packed column with various types of internals 填料塔示意图（带各种形式内件）
37　reboiler return 再沸器返回口
38　vapor injection packing support plate 气体喷射式填料支承板

39　random packing 散装填料
40　vapor feed 气体闪蒸进料
41　liquid feed 液体进料
42　ring channel with drainage 集液槽
43　support grid 填料支承栅板
44　structured packing 规整填料
45　locating grid 填料压圈
46　demister 除雾器，除沫器
47　reflux from condenser 回流入口
48　trough-type liquid distributor 槽式液体分布器
49　nozzle type liquid distributor 盘式液体分布器
50　vortex breaker 防涡流挡板
51　bottom product 塔釜出料口

Exercise

Write the name of the parts of the packing column.

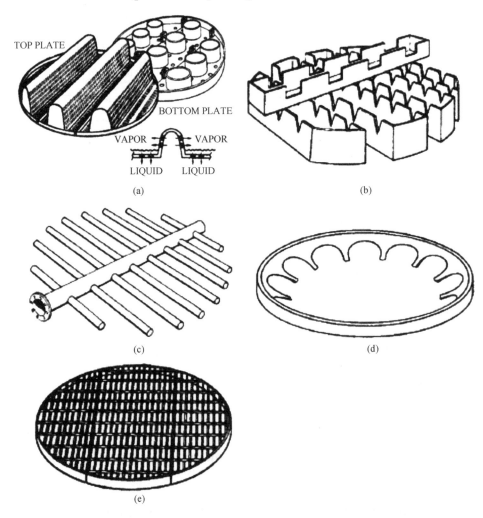

Text C Petrochemicals

Petroleum is perhaps the most important substance consumed in modern society. It provides not only raw materials for the ubiquitous plastics and other products, but also fuel for energy, industry, heating, and transportation. The word petroleum, derived from the Latin petra and oleum, means literally rock oil and refers to hydrocarbons that occur widely in the sedimentary rocks in the form of gases, liquids, semisolids, or solids.

Petroleum is a mixture of gaseous, liquid, and solid hydrocarbon compounds that occur in sedimentary rock deposits throughout the world and also contains small quantities of nitrogen-, oxygen-, and sulfur-containing compounds as well as trace amounts of metallic constituents.

Petroleum refining begins with the distillation or fractionation of crude oils into separate fractions of hydrocarbon groups. The resultant products are directly related to the characteristics of the crude oil being processed. Most of these products of distillation are further converted into more useable products by changing their physical and molecular structures through cracking, reforming, and other conversion processes. These products are subsequently subjected to various treatment and separation processes, such as extraction, hydrotreating, and sweetening, to produce finished products. Whereas the simplest refineries are usually limited to atmospheric and vacuum distillation, integrated refineries incorporate fractionation, conversion, treatment, and blending with lubricant, heavy fuels, and asphalt manufacturing; they may also include petrochemical processing. It is during the refining process that other products are also produced. These products include the gases dissolved in the crude oil that are released during distillation as well as the gases produced during the various refining processes that provide fodder for the petrochemical industry.

The gas (often referred to as refinery gas or process gas) varies in composition and volume, depending on the origin of the crude oil and on any additions (i.e., other crude oils blended into the

Words and expressions

petrochemical [ˈpɛtroˈkɛmɪkl] n. 石油化学品
petroleum [pəˈtrolɪəm] n. 石油
ubiquitous [juˈbɪkwɪtəs] adj. 普遍存在的
plastic [ˈplæstɪk] n. 塑料制品
fuel [ˈfjuəl] n. 燃料
latin [ˈlætn] n. 拉丁语
literally [ˈlɪtərəli] adv. 真正地
sedimentary [ˌsɛdɪˈmɛntri] adj. 沉积的
gaseous [ˈgæsɪəs] adj. 气态的
deposit [dɪˈpɑzɪt] n. 沉淀物
trace amount 痕量
constituent [kənˈstɪtʃuənt] n. 成分
crude oil 原油
characteristic [ˌkærəktəˈrɪstɪk] n. 特性，特征

cracking [ˈkrækɪŋ] n. 裂化
reforming [riˈfɔːmɪŋ] 重整
extraction [ɪkˈstrækʃən] n. 萃取
sweetening [ˈswitnɪŋ] n. 脱臭

integrate [ˈɪntɪgret] adj. 整体的
incorporate [ɪnˈkɔːpəreɪt] v. 包含
blend [blɛnd] v. 混合
lubricant [ˈlubrɪkənt] n. 润滑油
asphalt [ˈæsfɔlt] n. 沥青

fodder [ˈfɑdɚ] n. 素材

feedstock n. 进料

refinery feedstock) to the crude oil made at the loading point. It is not uncommon to reinject light hydrocarbons such as propane and butane into the crude oil before dispatch by tanker or pipeline. This results in a higher vapor pressure of the crude, but it allows one to increase the quantity of light products obtained at the refinery. As light ends in most petroleum markets command a premium, while in the oil field itself propane and butane may have to be re-injected or flared, the practice of spiking crude oil with liquefied petroleum gas is becoming fairly common. These gases are recovered by distillation. In addition to distillation, gases that are produced in the various thermal cracking processes are also available.

The relationship between raw materials and primary petrochemicals is shown in Fig. 2.7.

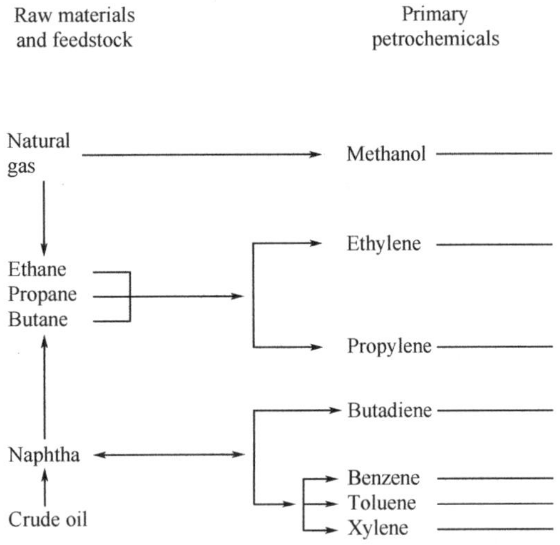

reinject [ˌriːɪnˈdʒɛkt] vt. 再注入
propane [ˈproʊpeɪn] n. 丙烷
butane [ˈbjuːten] n. 丁烷

premium [ˈprimɪəm] n. 附加费
flare [flɛr] vi. 燃烧
spiking [ˈspaɪkɪŋ] n. 突然上升，剧增
liquefy [ˈlɪkwɪfaɪ] v. 液化，溶解
thermal [ˈθɜːml] adj. 热的

Fig. 2.7 The relationship between raw materials and primary petrochemicals

Adapted from The Chemistry and Technology of Petroleum by Heinz Heinemann

Comprehension

Answer the following questions.

1. Is petroleum the same as crude oil?
2. Now petroleum has become the main source of chemical product. Explain it with examples.
3. What is the basic process to separate the crude oil into fractions?
4. Petroleum is a complex mixture. List its components.

Vocabulary building

Active words

deposit
n. 储蓄，存款，保证金；沉淀物；oil deposit 油田

subject
n. 主题；学科
adj. 须服从……的

origin
n. 起源，根源

convert
vt.（使）转变

reform
vt. 改良，改革，改变；[化学]重整

whereas
conj. 然而；鉴于；反之

release
vt. 释放

thermal
adj. 热的，保热的；温热的

Useful expressions

be derived from　来自
in the form of　用……形式
be related to　与……有关
not only…but also…　不仅……而且……

in addition to　除了……之外
as well as　又；也；此外
trace amount　痕量

Exercise

Translate the following sentences into English with the phrases in brackets.
1. 巧克力有痕量的咖啡因。(trace amount)
2. 板式塔和填料塔都可用于精馏操作。(as well as)
3. 石油精馏能分离出的组分与操作条件有关。(be related to)
4. 精馏和吸收操作的主要反应器都是塔设备。(not only…but also)
5. 石油来自于数亿万年前水生动物和植物的遗骸。(derived from)

Exercise

Fill in the blanks with the given words.

The vertical shell together with the condenser and reboiler constitute a distillation column. Let us consider how the column operates.

Assume, for simplicity, that the feed is a binary liquid mixture that is to be separated into two relatively pure products. Feed enters the central portion of the column on a tray called the feed tray. It flows _____ (up or down) by gravity from tray to tray and in the process comes into contact on each tray with the vapor rising from the tray below. The liquid from tray flows into the base of the column and then into a reboiler, where it is partially _____ (vaporize or condense). The unvaporized liquid is one of the products of the distillation operation. It is called the bottoms product, and is removed from the reboiler. The bottoms product has the _____ (highest or lowest) concentration of the least volatile substance and thus its temperature, which is also the temperature of the vapor generated in the reboiler, is the highest of any location in the column. Since the volatilities of the two substances involved in the distillation process are different, the vapor generated in the reboiler is richer in the _____ (more or less) volatile component. This vapor rises and comes into contact with the descending stream of liquid on each tray, beginning with that on tray. The mixing of warmer vapor with the liquid results in the transfer of heat and mass, and the net result is some vapor vaporization of the more volatile component and condensation of a thermally equivalent amount of the less volatile component. Thus the feed is "stripped" of its _____ (more or less) volatile component as it flows (upward or downward) and it becomes _____ (more or less) concentrated in the less volatile substance. The feed tray and the trays below it constitute what is called the stripping section. The feed tray and the trays above it constitute what is called the distillate section. The vapor rising from the feed tray comes into contact with a liquid that is more concentrated in the more volatile substance on each tray where some of the more volatile substance vaporizes at the expense of some of the less volatile substance, which condenses. Thus the vapor becomes "enriched" in the more volatile substance as it flows _____ (up or down) the column. The vapor from the top tray contains a higher concentration of the more volatile substance than anywhere else in the column. It is condensed in a total condenser, a part of it is removed as the distillate product, and the remainder is returned to the column as reflux. This liquid, which is the purest in terms of the more volatile component, contacts the ascending vapor rising from the feed tray and the heat and mass transfer processes occur as described earlier. The reflux stream combines with the feed and serves as the liquid phase in the stripping section.

Extension

Bitumen Blowing

Asphaltic bitumen, normally called "bitumen", is obtained by vacuum distillation or vacuum flashing of an atmospheric residue. This is "straight run" bitumen. An alternative method of bitumen production is by precipitation from residual fractions by propane or butane-solvent deasphalting.

The bitumen thus obtained has properties which derive from the type of crude oil processed and from the mode of operation in the vacuum unit or in the solvent deasphalting unit. The grade of the bitumen depends on the amount of volatile material that remains in the product: the smaller the amount of volatiles, the harder the residual bitumen.

In most cases, the refinery bitumen production by straight run vacuum distillation does not meet the market product quality requirements. Authorities and industrial users have formulated a variety of bitumen grades with often stringent quality specifications, such as narrow ranges for penetration and softening point. These special grades are manufactured by blowing air through the hot liquid bitumen in a bitumen blowing unit. What type of reactions take place when a certain bitumen is blown to grade? Bitumen may be regarded as colloidal system of highly condensed aromatic particles (asphaltenes) suspended in a continuous oil phase. By blowing, the asphaltenes are partially dehydrogenated (oxidized) and form larger chains of asphaltenic molecules via polymerization and condensation mechanism. Blowing will yield a harder and more brittle bitumen (lower penetration, higher softening point), not by stripping off lighter components but changing the asphaltenes phase of the bitumen. The bitumen blowing process is not always successful: a too soft feedstock cannot be blown to an on-specification harder grade.

The blowing process is carried out continuously in a blowing column. The liquid level in the blowing column is kept constant by means of an internal draw-off pipe. This makes it possible to set the air-to-feed ratio (and thus the product quality) by controlling both air supply and feed supply rate. The feed to the blowing unit (at approximately 2100℃), enters the column just below the liquid level and flows downward in the column and then upward through the draw-off pipe. Air is blown through the molten mass (280-3000℃) via an air distributor in the bottom of the column. The bitumen and air flow are countercurrent, so that air low in oxygen meets the fresh feed first. This, together with the mixing effect of the air bubbles jetting through the molten mass, will minimise the temperature effects of the exothermic oxidation reactions: local overheating and cracking of bituminous material. The blown bitumen is withdrawn continuously from the surge vessel under level control and pumped to storage through feed/product heat exchangers.

Exercise

Translate the following sentences into Chinese.

1. In most cases, the refinery bitumen production by straight run vacuum distillation does not meet the market product quality requirements. Authorities and industrial users have formulated a variety of bitumen grades with often stringent quality specifications, such as narrow ranges for penetration and softening point. These special grades are manufactured by blowing air through the hot liquid bitumen in a BITUMEN BLOWING UNIT.

2. The blowing process is carried out continuously in a blowing column. The liquid level in the blowing column is kept constant by means of an internal draw-off pipe. This makes it possible to set the air-to-feed ratio (and thus the product quality) by controlling both air supply and feed supply rate.

Answer the following questions according to the passage.

1. What factors does the grade of the bitumen depend on?
2. What reactant is the bitumen produced from?
3. How does the bitumen blowing carried out?
4. What is the bitumen blowing used for?

Reading material

Ethylene Production

Ethylene

Ethylene, $H_2C=CH_2$, is the lightest ole-fin. It is a colorless, flammable gas, which is produced mainly from petroleum-based feedstocks by thermal cracking in the presence of steam. Ethylene has almost no direct enduses but acts almost exclusively as an inter-mediate in the manufacture of other chemicals, especially plastics.

Ethylene may be polymerized directly to produce polyethylene, the world's most widely used plastic. Ethylene can also be chlorinated to produce 1, 2-dichloroethane, a precursor to the plastic polyvinyl chloride, or combined with benzene to produce ethylbenzene, which is used in the manufacture of polystyrene, another important plastic. Smaller amounts of ethylene are oxidized to produce chemicals including ethylene oxide, ethanol, and polyvinyl acetate.

Raw Materials

Various feedstocks (liquid and gaseous) are used for the production of ethylene. The principal feedstocks are naphtas, a mixture of hydrocarbons in the boiling range of 30℃ to 200℃. Depending on the origin, naphta composition and quality can vary over a wide range requiring quality control of the feed mixtures.

Production

The bulk of the worldwide production is based on thermal cracking with steam. The process is called pyrolysis or steam cracking. Production can be split into four sections (Fig. 2.8):

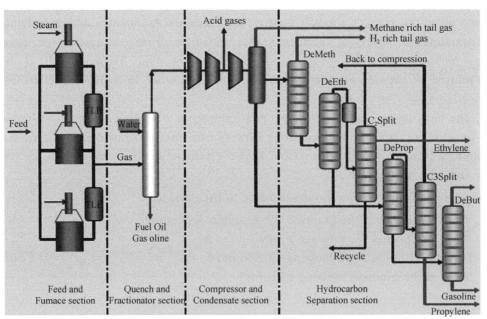

Fig. 2.8 Ethylene production (overview)

The first three sections are more or less identical for all commercial processes, with the exception that primary fractionation is required only in case of a liquid feedstock.

A large variety of process routes, however, exist for the hydrocarbon fractionation section. A hydrocarbon feed stream is preheated, mixed with steam and further heated to 500℃ to 700℃. The stream enters a fired tubular reactor (known as cracker, cracking heater), where under controlled conditions the feedstock is cracked at 800℃ to 850℃ into smaller molecules within a residence time of 0.1s to 0.5 s. After leaving the radiant coils of the furnace the product mixtures are cooled down instantaneously in transfer line exchangers (TLE) to preserve the gas composition. This quenching time is a crucial measure for severity control of the final products.

The steam dilution lowers the hydrocarbon pressure, thereby enhancing the olefin yield and reducing the tendency to form and deposit coke in the tubes of the furnace and coolers.

Cracking furnaces (capacity of modern units up to 150 000 t/year) represent the largest energy consumer in an ethylene plant. Cracking furnace technologies are offered by engineering companies such as ABB Lummus, KTI-Technip, Linde AG (Pyrocrack), M. W. Kellog, Stone & Webster, e. a.

Other processes for ethylene production besides conventional thermal cracking include:
- Recovery from Fluid Catalytic Cracking (FFC) offgas
- Fluidized-bed cracking
- Catalytic pyrolysis
- Membrane reactor
- e. a.

Process Optimization

Process optimization is critical for ethylene production because cracking reactions change as the run proceeds. Operation costs are high and, therefore, process control including online analyzers providing almost realtime process information has reached a very high level of importance. Models for different kinds of feedstocks have been developed to optimize production of certain amounts of ethylene, propylene and other products at maximum profit even with changing of feedstock quality or type.

Adapted from Process Analytics in Ethylene Production Plants@Siemens AG 2007—
www. siemens. com / processanalytics

Comprehension

Answer the following questions according to the passage.
1. What are the main usages of ethylene?
2. What are the chemical properties of ethylene?
3. How many sections may the production of ethylene include?
4. What are the byproducts of ethylene production?

Supplementary knowledge

Structure of Equipment Manual

A user guide or user's guide, also commonly known as a manual, is a technical communication document intended to give assistance to people using a particular system. User guides are most commonly associated with electronic goods, computer hardware and software.

The sections of a user manual often include:
- a cover page (see Fig. 2.9);

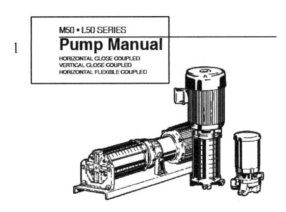

Fig. 2.9 The title page

1—the equipment's name; 2—the publication time and edition; 3—the contact details

- the general instructions;
- the installation (see Fig. 2.10);

M50 · L50 SERIES
2. Installation

FLEXIBLE COUPLED PUMPS
CLOSE COUPLED PUMPS

A. Location
B. Foundation
C. Leveling
D. Alignment

E. Piping
F. Typical Installation

In order to insure that pumping equipment is installed properly and to obtain reliable pump operation, it is recommended that only experienced, qualified erecting engineers undertake this task. Read the instructions thoroughly before beginning.

2A Location

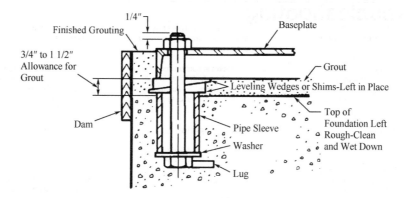

Fig. 2.10 Foundation

- the operation (see Fig. 2.11);

M50 · L50 SERIES
3. Operation
FLEXIBLE COUPLED PUMPS
CLOSE COUPLED PUMPS

A. Rotation
B. Inlet and Outlet Locations
C. Foreign Material
D. Electrical
E. Adjustments
F. Cooling Water
G. Priming
H. Starting
I. Stopping

3A Rotation

The standard direction of rotation for the pump is right-handed, or clockwise when looking at the motor end of the pump. A rotation arrow, refer to Figure 3-1, is located on the pump to indicate the correct direction of rotation.

Operating the pump in reverse will cause substantial performance variations and can damage the pump.

Always confirm correct motor rotation prior to connection of the coupling. If this is not possible, perform a final rotation check as follows:

1. Jog the motor briefly.
2. Observe rotation as the unit comes to a stop.
3. Rotation should be in the direction of the arrow.

If the motor operates in the wrong direction:

1. Interchange any two leads on a three-phase motor.
2. On a single-phase motor, change the leads as indicated on the connection box cover. Some single-phase motors may not be reversible.

3B Inlet and Outlet Locations
(Refer to Figure 3-1)

Fig. 2.11 Operation

- the service;
- troubleshooting (see Fig. 2.12); detailing possible errors or problems that may occur, along with how to fix them;
- repair services.

M50 · L50 SERIES

5. Troubleshooting

FLEXIBLE COUPLED PUMPS
CLOSE COUPLED PUMPS

A. Failure to Pump
B. Reduced Capacity
C. Reduced Pressure
D. Pump Loses Prime After Starting
E. Excessive Power Consumption
F. Pump Vibrates or is Noisy
G. Mechanical Problems
H. Seal Leakage

5A Failure to Pump

1. Pump not up to speed — Use Tachometer to determine actual RPM. Check voltage and wiring connections.
2. Pump not primed — Confirm that pump and all inlet piping are filled with fluid.
3. Discharge head too high — Install a pressure gauge at the pump discharge to determine the actual operating pressure. Compare readings with pump performance curve. A larger pump may be necessary.
4. Excessive suction lift — Relocate pump, supply tank, or both to minimize suction lift.
5. Wrong direction of rotation — Compare pump rotation with arrow on pump. Standard pumps rotate in a clockwise direction when looking at the shaft extension end or from the motor end on close-coupled pumps. Reverse two leads on a three-phase motor to change rotation. Check motor nameplate for single-phase operation.
6. Clogged suction line, strainer, or foot valve — Inspect and clean out if necessary.
7. Air pocket in suction line — Look for high spots in inlet piping system. Evacuate the system with a vacuum pump if necessary.

5B Reduced Capacity

1. Pump not up to speed — Use a tachometer to determine actual RPM. Check voltage and wiring connections.
2. Excessive suction lift — Relocate pump, supply tank, or both to minimize suction lift.
3. Insufficient NPSH — Relocate pump, supply tank, or both to improve NPSH available if possible. Increase suction pressure. Reduce fluid temperature. Select a pump with lower NPSH requirements.
4. Mechanical damage — Rotate the pump by hand to determine if there are tight spots. Broken or bent impeller vanes can sometimes be detected in this manner. If there is a suspicion of damage, remove the pump from service and disassemble for inspection.
5. Air leak in the suction line — Fill the system with fluid and hydrostatically test. Tighten connections or replace leaky components.
6. Air pockets in the suction piping — Operating the system at maximum flow conditions will usually clear the lines. Evacuate

Final Adjustments.

the system with a vacuum pump if necessary.

7. Suction lines, strainer, or foot valve too small or clogged — Inspect and clean out as necessary. Fittings and lines should be at least equal to the pump suction size.
8. Discharge head too high — Install a pressure gauge at the pump discharge to determine the actual operating pressure. Compare readings with pump performance curve. A larger pump may be necessary.
9. Excessive wear — If a pump had previously performed satisfactorily and now gives evidence of reduced performance, it should be disassembled and examined for wear after the simpler possible problems have been investigated.

5C Reduced Pressure

1. Pump not up to speed — Use a tachometer to determine actual RPM. Check voltage and wiring connections.
2. Air or vapor in liquid — Install a separator in the suction line. Check the seal on the inlet end of the pump to determine if air is being drawn in. Hydrostatically test the system to insure that there are no leaks.
3. Mechanical wear or damage — Rotate the pump by hand to determine if there are tight spots. Broken or bent impeller vanes can sometimes be detected in this manner. If there is a

Fig. 2.12　Troubleshooting

Adapted from M50. L50 SERIES Pump Manual —— www.mthpumps.com

Exercise

Answer the following questions.

1. What is the function of a manual?
2. If the equipment is out of order, where can we get help to solve the problem?
3. Is the equipment manual published?
4. How many procedures are needed to install a pump and what are they?

Chemical Engineering Safety

Objectives:

After finishing this module, you are able to:
- Recognize hazardous chemicals signs
- Use the fire Extinguisher correctly
- Make precautions beforehand, address the incident at the beginning and in process of it, and deal with the aftermath
- Write simple technical & safety instructions for chemicals

Warming-up

What Do These Public Signs Mean?

Text A General rules for classification and hazard communication of chemicals

National standard of the People's Republic of China (GB 13690—2009) is issued by General Administration of Quality Supervision, Inspection and Quarantine of the People's Republic of China and Standardization Administration of the People's Republic of China. Chapter VI and chapter V in the Standard are mandatory, while the rest are recommended.

The conformity degree between this Standard and its corresponding regulation, the second revised edition of "*Globally Harmonized System of Classification and*

Words and expressions

classification [ˌklæsɪfɪˈkeɪʃən] n. 分类
hazard [ˈhæzəd] n. 危险
standard [ˈstændəd] n. 标准
issue [ˈɪʃuː] vt. 发布
mandatory [ˈmændətɔri] adj. 强制的

revised [rɪˈvaɪzd] v. 修改
conformity [kənˈfɔrməti] n. 符合

Labeling of Chemicals" (GHS) is non-equivalent. This Standard's technical contents are in conformity with those of the GHS. In accordance with GB/T 1.1—2000, some editorial changes have been made to the Standard's format. The following is part of the Standard.

4. Classification

4.1　Physical and chemical hazards

4.1.1　Explosive

Please refer to GB 20576 for the classification, warning labels and warning statements for explosive.

refer to 参考；涉及

4.1.1.1

An explosive substance (or mixture) is a solid or liquid which is in itself capable by chemical reaction of producing gas at such a temperature and pressure and at such a speed as to cause damage to the surroundings. Pyrotechnic substances are included even when they do not evolve gases.

A pyrotechnic substance (or mixture) is designed to produce an effect by heat, light, sound, gas or smoke or a combination of these as the result of non-detonative, self-sustaining, exothermic chemical reactions.

An explosive article is a material that contains one kind or several kinds of explosive substances or mixtures.

A pyrotechnic article is a material that contains one kind or several kinds of pyrotechnic substances or mixtures.

surroundings [səˈraʊndɪŋz] n. 环境
pyrotechnic [ˌpaɪrəˈtɛknɪks] adj. 烟火的
combination [ˌkɑmbɪˈneʃən] n. 结合
exothermic [ˌɛksoˈθɜmɪk] adj. 放热的
explosive [ɪkˈsplosɪv] adj. 爆炸的

4.1.1.2

Classification of explosives comprises：

a) Explosive substances and mixtures.

b) Explosive articles. however, the following devices are excluded: the explosive substances or mixtures contained in the devices, due to the quantity or property of the substances or mixtures, are unable to produce any effect outside the devices upon accidental or occasional ignition after emission, fire, smoke, heat or a loud booming sound occurs.

c) Substance, mixtures and articles, not mentioned in a) and b), that are designed to produce actual explosion or firework effect.

comprise [kəmˈpraɪz] vt. 包含；由……组成

ignition [ɪgˈnɪʃən] n. 点燃 emission [ɪˈmɪʃən] n.（光、热等的）发射，散发
boom [bʊm] vt. 发隆隆声

4.1.2　Flammable Gases

Please refer to GB 20577 for the classification, warning labels and warning statements for flammable gases.

Flammable gas means a gas having a flammable

flammable [ˈflæməbl] adj. 易燃的

range in air at 20℃ and a standard pressure of 101.3kPa.

4.1.3 Flammable Aerosols

Please refer to GB 20578 for the classification, warning labels and warning statements for flammable aerosols.

Aerosols are any gas compressed, liquefied or dissolved under pressure within a non-refillable container made of metal, glass or plastic, with or without a liquid, paste or powder. The container is fitted with a release device allowing the contents to be ejected as solid or liquid particles in suspension in a gas, as a foam, paste or powder or in a liquid or gaseous state.

aerosol ['ɛrəsɔl] n. 气溶胶

eject [ɪ'dʒɛkt] vt. 喷射
suspension [sə'spɛnʃən] n. 悬浮
paste [pest] n. 糊状物

Comprehension

Choose the best answer according to the text.

1. Which one does not belong to flammable gas?
   A. methane   B. hydrogen sulfide
   C. propane   D. carbon dioxide

2. Flammable gas means a gas having a flammable range in air at (　　) and a standard pressure of (　　).
   A. 20℃　101.3kPa   B. 25℃　101.3kPa
   C. 20℃　1.013kPa   D. 25℃　1.013kPa

3. A/An (　　) is a solid or liquid which is in itself capable by chemical reaction of producing gas at such a temperature and pressure and at such a speed as to cause damage to the surroundings.
   A. explosive articles   B. flammable gas
   C. explosive substance   D. aerosols

4. What will be discussed in the next section?
   A. health hazards   B. environmental hazards
   C. hazard communication D. hazard category

Vocabulary building

Active words

classification
　n. 分类；类别

explosive
　adj. 爆炸的
　n. 爆炸，爆炸物

flammable
　adj. 易燃的
　n. 易燃物

emission
　n. 排放；发射

hazard
　n. 危险；危险品

article
　n. 条款；文章

surrounding
　adj. 周围的

n. 环境

evolve

v. 进化；释放

Useful expressions

in accordance with 与……一致
refer to 引证；查阅，参考
as the result of 由于

issue

vt. 发行；放出

due to 由于
conformity with 与……一致

Exercise

Fill in each blank with a given word or expression in their right form.

explode   revise   flame   issue   exothermic   comprise

1. The powder keg was an _____ matter.
2. A _____ liquid, C_2H_5CHO, are used in the manufacture of plastics and rubber chemicals.
3. The standard for the matter has been completely _____.
4. Without cooling water, the irradiated nuclear fuel could spontaneously combust in a/an _____ reaction.
5. Credit cards are _____ by many large department stores to their customers.
6. The ideal post-workout meal is _____ of a blend of carbohydrates and protein.

Extension

P. A. S. S. for Extinguisher

P. A. S. S. is an acronym used to describe the actions taken to properly use a halon, dry chemical or carbon dioxide fire extinguisher (see Fig. 3.1). It stands for Pull, Aim, Squeeze and Sweep, the four steps to successful fire fighting.

● Pull. "P" stands for "pull". and it refers to the locking pin on the handle. The pin is used to prevent the fire extinguisher from being discharged accidentally. In order for the lever to work, the pin must be removed.

● Aim. "A" stands for "aim". If the fire extinguisher discharges from the end of a hose, point the hose at the base of the fire. If your extinguisher discharges from a nozzle at the top of the canister, point it directly at the base of the fire. P. A. S. S. -style fire extinguishers discharge only for about 10 seconds, so it's important to aim first so you don't waste your extinguisher.

Fig. 3.1   Extinguisher

● Squeeze. The first "S" in P. A. S. S. stands for "squeeze". Once your extinguisher is properly aimed, pull the lever upward toward your palm.

● Sweep. Sweep the nozzle side-to-side toward the base of the fire to make sure all potential hotspots are saturated. Don't hesitate to use the extinguisher until it is empty.

Text B　Technical & Safety Instructions for Chemicals

Section 1　Identifications of the Chemical and the Enterprise

Chinese Name of the Chemical：液氨
English Name of the Chemical：Liquid ammonia
Enterprise Name：Refinery of Jinling Petrochemical Corp., Ltd.

Section 2　Component and Composition Information

pure substance：☑　　　　mixture：☐
Name of the product：Liquid ammonia
The information of ammonia is shown in table 3.1.

Table 3.1　Composition information

Noxious Substance	Concentration	CAS NO.
ammonia	≥99.6%	7664—41—7

Section 3　Hazard Description

Category of Hazard：Toxic gas, category 2.3
Ways of Incursion：Inhalation
Health Hazard：Ammonia with low concentration has an irritant effect on mucous membrane, and ammonia with high concentration may cause histolysis necrosis. Environment Hazard：The product may seriously contaminate the environment, in particular surface water, soil, air and drinking water.

Fire and Explosion Hazard：The product can form an explosive mixture with air, and combustion and explosion may occur when it comes across naked flame or at high temperature. The product can vigorously react with fluorine and chlorine, etc., when contacting them. The container for the product may crack or explode at high temperature because of the increase of pressure.

Section 4　First Aid Measures

Skin Contact：Take off the contaminated clothes immediately, flush the injured part of skin with 2% boracic acid liquor or large amount of clean water. Send the injured person to hospital.

Spattering into eyes：Lift eyelids immediately, and wash the eyes with large amount of flowing clean water or normal saline thoroughly for at least

Words and expressions

identification [aɪˌdɛntəfɪ'keʃən] n. 鉴定，识别

ammonia [ə'monɪə] n. 氨
refinery [rɪ'faɪnəri] n. 精炼厂

component [kəm'ponənt] n. 成分
composition [ˌkɑmpə'zɪʃən] n. 构成

CAS [ˌsie'ɛs] CAS 号；化学物质登记号

inhalation [ˌɪnhə'leʃən] n. 吸入
mucous ['mjukəs] adj. 黏液的；分泌黏液的
membrane ['mɛmbren] n. 膜

combustion [kəm'bʌstʃən] n. 燃烧
fluorine ['flu(:)əri:n] n. 氟（元素符号 F）
chlorine ['klɔrin] n. 氯（17 号化学元素）

contaminate [kən'tæmɪnet] vt. 污染
injure ['ɪndʒɚ] vt. 伤害，损害

15 minutes. Send the injured person to hospital.

Inhalation: Quickly remove the injured person from the polluted locale to the place with fresh air, and keep the respiratory tract expedite. If it is difficult for the injured person to breathe, treat him or her with oxygen therapy. If the injured person ceases to breathe, immediately treat him or her with artificial respiration. Send the injured person to hospital.

Ingestion: No data available.

Section 5 Operation and Storage

Operation Precautions: The product must be kept far from tinder and heat sources. The storehouse must be prevented from direct solarization. Mechanical equipment or tools that may produce sparkles easily mustn't be used.

tinder ['tɪndɚ] n. 易燃物
solarization [ˌsolərɪ'zeʃən] n. 日晒

Storage Precautions: The product must be stored in cool, dry and ventilated storehouse. The containers of the product must be separately stored, apart from halogens (fluorine, chlorine and bromine), acids, etc.

precaution [prɪ'kɔʃen] n. 预防措施

Section 6 Contact Control and Safeguards for Individual

Permissible Maximum Concentration: China MAC (mg/m^3), 30

safeguard ['sefgɑrd] n. 保护措施

Monitoring Technique: Nessler's reagent colorimetry Safeguards for Respiration: It is recommended that filtrating gas masks be worn when the product concentration in air exceeds specified value. Air respirators must be worn in the case of emergency rescue or withdrawal.

Safeguards for Eyes: Wear chemical proof safeguard glasses.

Safeguards for Bodies: Wear static electricity proof suits.

Safeguards for Hands: Wear rubber gloves.

Section 7 Treatment for the Waste

Property of the Waste: Hazardous waste

Treatment for the Waste: The waste should be treated in line with relevant national and local laws and regulations. It must be incinerated or processed with biodegradation methods.

in line with 符合；与……一致

Comprehension

Choose the right answer according to the text.

1. According to the passage, which of the following statement is correct?
   A. Liquid ammonia is an elementary substance.
   B. Liquid ammonia is a mixture.
   C. Liquid ammonia is a pure substance.
   D. Liquid ammonia is composed of compounds.
2. Which is not the way of toxic matter entering human body?
   A. inhalation   B. injection   C. ingestion   D. skin contact
3. Which of the following is NOT the reason why liquid ammonia explodes?
   A. Naked flame   B. High temperature   C. Sparkle   D. Vacuum
4. Which one is the name of NH_3?
   A. ammonia   B. amine   C. ammonium hydroxide   D. amide

Write down the contact control and safeguards for individual in English.

Safeguards for Respiration:
_____
_____

Safeguards for Eyes:
_____
_____

Safeguards for Bodies:
_____
_____

Safeguards for Hands:
_____
_____

Vocabulary building

Active words

contaminate
vt. 污染，弄脏

injure
vt. 伤害，损害

precaution
n. 预防措施

safeguard
n. [安全] 保护措施

Useful expressions

in case of   在……的情况下
in line with   符合；与……一致

apart from   除……之外

Exercise

Fill in each blank with a given word or expression in their right form.

wear   refine   contaminate   injure   contain   identify

1. Dalian is one of China's major oil production and distribution centers, and there is one of the nation's largest _____.
2. Please _____ chemical proof safeguard glasses.
3. The nurse bandaged up his _____ finger.
4. He drained all the old oil out of the _____.
5. Methods: To analyze and _____ the chemical constituents by GC-MS.
6. The river was _____ with waste from the factory.

Extension

How to do CPR?

The CPR (cardiopulmonary rescuscitation) is shown in Fig 3.2.
1. Place the victim on his or her back.

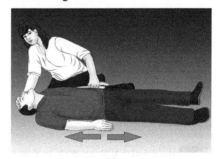
(a)

2. Place the heel of one hand on the victim's breastbone.

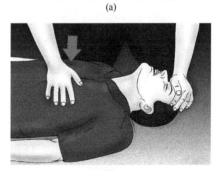
(b)

3. Place your second hand on top of the first hand.

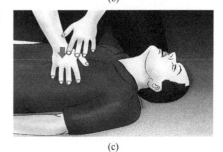
(c)
Fig. 3.2

4. Position your body directly over your hands.

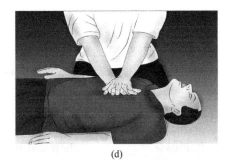
(d)

5. Perform 30 chest compressions.

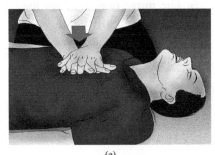
(e)

6. Continuing the process until help arrives.

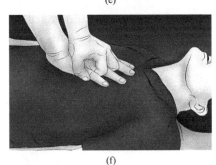
(f)

7. Make sure the airway is open.

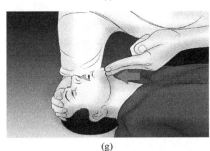
(g)

8. Give two rescue breaths (optional).

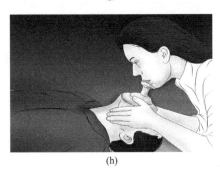
(h)

Fig 3.2 The CPR
Adapted from www. wikihow. com

Text C Accident Investigation Report

Executive Summary

Sinopec NCIC (Nanjing Chemical Industry Company) Gasification Plant experienced a slurry pipe rupture on February 13rd, 2008. Based on the information received from Sinopec, GEENERGY (USA), LLC (GE) believes that the loss of containment was caused by oxygen and syngas backflow into the slurry line when an operator mistakenly opened a drain valve on the operating slurry train.

Sinopec invited GE to attend a meeting with NCIC and Jinling (NCIC's sister plant) to analyze the incident and provide recommendations on improvements needed to prevent similar incidents from reoccurring in the future. The meeting took place on March 11st, 2008 at Sinopec's corporate office in Nanjing City, China. Sinopec NCIC provided GE with copies of DCS screen shots showing trends of critical process parameters and the sequence of events that took place during the incident. The documents received can be found in appendix A of this report. GE requested permission to visit the plant site during the review meeting on March 11st, 2008, but was denied access.

GE's assessment of this incident is based on GE's interpretation of the information received during the review meeting with Sinopec on March 11st, 2008, and the DCS screen shots received subsequently. Eyewitnesses to the incident were not interviewed during the investigation, and Sinopec did not provide GE with a written transcript of the incident for the investigation.

Based on the information received, GE believes that human errors and inadequate instrument response are two categories of root causes that may have led to this incident. Proposed in this report are a number of operational and process modifications that can be implemented to address the areas of concern identified. Shown below is a summary of the root causes

Words and expressions

investigation [ɪnˌvɛstɪˈgeʃən] n. 调查；调查研究

rupture [ˈrʌptʃɚ] vi. 破裂

plant [plænt] n. 工厂

syngas [ˈsingæs] n. 合成气

valve [vælv] n. 阀

recommendation [ˌrɛkəmɛnˈdeʃən] n. 推荐；建议

parameter [pəˈræmətɚ] n. 参数

appendix [əˈpɛndɪks] n. 附录

assessment [əˈsɛsmənt] n. 评定；估价

eyewitness [ˈaɪwɪtnəs] n. 目击者；见证人

modification [ˌmɑdɪfɪˈkeʃən] n. 修改，修正

and risk mitigation actions recommended in the report.

Root Causes
- Mis-operation of slurry drain valve
- Slow response of slurry magnetic flow meter
- Incorrect setting of magnetic flow meter

Substitution of 3rd magnetic flow meter with Slurry Pump speed flow calculator

Recommendations
- Improvements to plant operations policies: fail-safe LOTO, tiered PTW, valve tagging and SOP's
- Addition of slurry check valve
- Addition of new shutdown initiators to ESS
- Use of triple-redundant slurry magnetic flow meters
- Modification of magnetic flow meter reverse flow settings
- General maintenance guidelines around slurry feed system
- Installation of a plant data collection system

The effectiveness of some of the process modifications proposed in this report will need to be verified through field-testing and data gathering at the plant site. GE is available to assist Sinopec with the efforts required in testing and validating these modifications. GE is also available to provide opera-tions training on the operations management mitigation actions recommended in the report. GE recommends that Sinopec should review the current conditions at all three of its plants (NCIC, Jingling and Qilu) against the recommendations and findings in this report.

Incident Description
Incident Narrative:
Sinopec NCIC (Nanjing Chemical Industry Company) Gasification Plant experienced a slurry pipe rupture during the morning hours of February 13_{rd}, 2008. Gasifier Train B was in operation and Train A was on stand by. The plant planned to start Train A. The field operator was instructed to open the drain valve at the outlet of Train A Slurry Pump in order to flush the pump. The operator made a

mitigation [ˌmɪtɪˈgeʃən] n. 减轻；缓和；平静
recommend [ˈrɛkəmɛnd] vt. 推荐

magnetic flow meter 电磁流量计

initiator [ɪˈnɪʃɪetɚ] n. 发起人

field-test [ˈfiːld-test] vt. 现场测试
assist…with 帮助（照料，做）
在……给予帮助
validate [ˈvælɪdet] v. 确认

gasifier [ˈgæsifaiə] n. 气化炉

mistake and opened the drain valve for Train B pump instead, causing the slurry line to depressurize immediately. The depressurization of the slurry line caused the high-temperature, high-pressure, syngas and oxygen in the Gasifier (80barg operating pressure) to backflow into the slurry line. This resulted in a loss of containment of the length of slurry pipe between the Slurry Charge Pump and slurry recycle line. The Slurry Pump status signal went to STOP and the Gasifier was tripped about 55 seconds after the loss of containment. Sinopec reported that the wiring of the pump was damaged during the incident, and that probably caused the pump status to go to STOP and trip the Gasifier.

DCS Events Tracing:

Sinopec reported that all the shutdown initiators (SDI's) for the slurry feed system including the Slurry Flow Low, O/C Ratio High and O/C Ratio High-High were armed during the incident. However, none of those initiators were triggered prior to the incident. Based on DCS screen shots provided to GE, the following is the timeline of events that took place during the incident:

At 08:57.00, the Gasifier pressure PT-52204 started to drop; it is assumed that the drain valve was opened at about that time. Refer to DCS Trend 1, Figure A-1, in appendix A.

• There were no significant changes in slurry and oxygen flows until around 08:57.15 at which point both readings began to rise.

• At 08:57.52, the slurry flow trend was lost, a potential indication that the flow meters were no longer functional.

• At 08:57.55, the oxygen flow reading began to drop; it is assumed that the Gasifier was manually tripped at about that time.

The magnetic flow meters (Krohne 4300 measure element with IFC 300 signal converter/ transmitter) had two unusual responses during this time period.

a) Slow response to flow change: Based on the DCS trends, there was no change in slurry flow for

depressurize [di'preʃə,raɪz] vt. 使减压

trigger ['trɪgɚ] vt. 引发, 引起; 触发

the first 15 seconds after the drain valve was believed to have been opened. From the SOE, two Low Flow alarms were recorded at different moments, FAXLL52201A at 08:57.37 and FAXLL 52201B at 08:57.40. The 2oo3❶ low slurry flow SDI was never triggered. GE was informed that the slurry pump flow calculator was being used as a substitute for the third magnetic flow meter during the time of the incident.

b) Unexpected increase in flow reading: The slurry flow meters' readings started increasing, not decreasing, around 15 seconds after the drain valve was believed to have been opened. Sinopec informed GE that the type of meter used, Krohne IFC 4300, has the capability of measuring reverse flow depending on the setting of the transmitter, and that the unexpected increase in slurry flow reading may have been due to the detection of reverse flow by the transmitter. After the incident, Sinopec modified the meters to indicate Zero upon the detection of reverse flow.

The O/C Ratio High trip and O/C Ratio High-High trip were never activated during the incident. The trip set point for O/C Ratio High trip with 10-second is 1.4 and the one for O/C Ratio High-High trip is 1.5 per NCIC's inputs. From the DCS screen shot Figure A-2 in appendix A, it appears that the last good value for O/C was 1.38731, very close to but not at the trip point before the loss of containment.

Comprehension

Answer the following questions.

1. Where and when did the accident happen?
2. What is the root causes of the accident?
3. How to prevent similar incidents?
4. Can you describe the incident in your own words?

---

❶ 2oo3, two out of three voting logic.

Vocabulary building

Active words

investigation
n. 调查

screen
n. 屏，幕
v. 筛选

syngas
n. 合成气

assessment
n. 评估

eyewitness
n. 目击者

trend
n. 趋势

trace
vt. 追踪

trigger
vt. 引发

assume
vt. 呈现

functional
adj. 功能的

substitute
n. 代用品

reverse
vi. 倒退

Useful expressions

magnetic flow meter 电磁流量计
lead to 导致

assist with 帮助
prior to 在……之前

Exercise

Fill in each blank with a given word or expression in their right form.

screen   function   assess   trace   assume   substitute

1. Recently, certain achievements have been made in the application of nano-technology in the development of _____ chemical fibers.
2. But how do you really know what exists unless you do an in-depth _____?
3. It's okay to _____ vegetable oil for butter.
4. This _____ was made on canvas.
5. We could not reason out which way the robbers escaped, because we were unable to find any _____ of them.
6. The problem has _____ a new form.

Translate the material safety data sheet of gasoil into Chinese.

Material Safety Data Sheet of Gasoil

1. First aid measures

Eye contact: Flush thoroughly with water. If irritation occurs, call a physician.

Skin contact: Remove contaminated clothing. Dry wipe exposed skin and clean yourself with waterless hand cleaner and follow by washing thoroughly with soap and water. For those providing assistance, avoid further contact to yourself or others. Wear impervious gloves. Launder contaminated clothing separately before reuse. Discard contaminated articles that cannot be laundered. (See Section 16 - Injection Injury)

Inhalation: Remove from further exposure. If respiratory irritation, dizziness, nausea, or unconsciousness occurs, seek immediate medical assistance. If breathing has stopped, assist ventilation with mechanical device or use mouth-to-mouth resuscitation.

Ingestion: Seek immediate medical attention. Do not induce vomiting.

Note to physicians: Material if aspirated into the lungs may cause chemical pneumonitis.

Pre-existing medical conditions which may be aggravated by exposure: Hydrocarbon Solvents/Petroleum Hydrocarbons-Skin contact may aggravate an existing dermatitis.

2. Fire-fighting measures

Extinguishing media: carbon dioxide, foam, dry chemical and water fog.

Special firefighting procedures: Water may be ineffective, but water should be used to keep fire-exposed containers cool. Prevent runoff from fire control or dilution from entering streams, sewers, or drinking water supply.

Special protective equipment: For fires in enclosed areas, fire fighters must use self-contained breathing apparatus.

Unusual fire and explosion hazards: Material is combustible. Liquid can release vapors that readily form flammable mixtures at or above the flash point. Product can accumulate a static charge which may cause a fire or explosion.

Combustion products: Fumes, smoke, carbon monoxide, sulfur oxides, aldehydes and other decomposition products, in the case of incomplete combustion.

Flash Point ℃ (℉): > 55 (131)(ASTM D-93).

Flammable Limits (approx. % vol. in air): LEL 0.6%, UEL 7.0%.

Nfpa hazard id: Health: 1, Flammability: 2, Reactivity: 0.

Extension

Accident Investigation Procedures

The actual procedures used in a particular investigation depend on the nature and results of the accident. The agency having jurisdiction over the location determines the administrative procedures. In general, responsible officials will appoint an individual to be in charge of the investigation. An accident investigator should use most of the following

steps:
- Define the scope of the investigation.
- Select the investigators. Assign specific tasks to each (preferably in writing).
- Present a preliminary briefing to the investigation team.
- Visit and inspect the accident site to get updated information.
- Interview each victim and witness. Also interview those who were present before the accident and those who arrived at the site shortly after the accident. Keep accurate records of each interview. Use a tape recorder if desired and if approved.

Determine the following:
- What was not normal before the accident.
- Where the abnormality occurred.
- When it was first noted.
- How it occurred.

Determine the following:
- Why the accident occurred.
- A likely sequence of events and probable causes (direct, indirect, basic).
- Alternative sequences.
- Determine the most likely sequence of events and the most probable causes.
- Conduct a post-investigation briefing.
- Prepare a summary report including the recommended actions to prevent a recurrence. Distribute the report according to applicable instructions.

An investigation is not complete until all data are analyzed and a final report is completed. In practice, the investigative work, data analysis, and report preparation proceed simultaneously over much of the time spent on the investigation.

Reading material

How to Use a Fire Extinguisher?

There is a high chance of your encountering an out-of-control fire at least once in your life, so knowing how to use a fire extinguisher is an important skill to be able to resort to. This article explains the process of using a fire extinguisher in an emergency (see Fig. 3.3).

Fig. 3.3 Using a fire extinguisher

1. Call for help before attempting to extinguish a serious fire. The fire may take hold much faster than you're capable of dealing with it, and if help is on the way, it removes one less concern for you.

- Call, or have someone else call, 119 in China (or the appropriate emergency for your country) as soon as possible. Ask for the fire service to come immediately, giving your address and a brief description of the type of fire.

Check that all other people are out of the house and have them remove pets as well. Check that they're all assembled at a safe meeting point. Do not allow children to attempt to use a fire extinguisher or control a fire in any respect whatsoever.

- Realize that reacting to a fire requires a sound process of decision-making that children and some adults may not be capable of coping with their panic. Prior training on correct use can alleviate some of the concern here.

2. Check for your own safety before starting to extinguish a fire. There are some key things to check for before you start fighting a fire using a fire extinguisher:

- Are you physically capable of extinguishing a fire? Some people have physical limitations that might diminish or eliminate their ability to properly use a fire extinguisher. People with disabilities, older adults, or children may find that an extinguisher is too heavy to handle or it may be too difficult for them to exert the necessary pressure to operate the extinguisher.

- Look for your exit points. Ensure that there is a clear exit for immediate escape should this becomes necessary. At all times, keep your mind focused on the availability of a safe retreat. If this is threatened at all, leave at once. The US National Fire Prevention Association recommends that you install fire extinguishers close to an exit point, to enable you to keep your back to the exit when you use the extinguisher; this ensures that you can make an easy escape if the fire cannot be controlled.

- Do not attempt to put out a fire where it is emitting toxic smoke; if you suspect or simply don't know if the smoke is toxic, leave it to the professionals.

- Check for structural safety of the building, in case burning walls, floors, or rafters pose a risk to your safety.

- If you have more than one fire extinguisher, consider asking another mature and responsible person to use it in tandem with you.

- Remember that your life is more important than property, so don't place yourself or others at risk.

3. Assess the fire. Only a contained fire should be fought using a fire extinguisher. Portable fire extinguishers are valuable for immediate use on small fires because they contain a limited amount of extinguishing material, which needs to be used properly or it will be wasted. For example, when a pan initially catches fire, it may be safe to turn off the burner, place a lid on the pan, and use an extinguisher. By the time the fire has spread, however, these actions will not be adequate, and only trained firefighters can safely extinguish such fires.

- Make a quick commonsense assessment about the utility and safety of using a fire extinguisher for the fire you're experiencing. Obviously, a fire extinguisher is overkill for a candle but it's useless when the whole house is on fire. A fire in a wastepaper basket, however, is another suitable candidate for fire extinguisher use.

• Use your instincts. If your instincts tell you the fire's too dangerous to tackle, trust them.

4. Check the type of extinguisher (see Fig. 3. 4). In the USA, there are five main classes of fire extinguisher (note that the classes and nomenclature can vary from country to country): A, B, C, and the less common classes D and K. The extinguishing agent might be water, dry chemical, halon, CO_2, or a special powder.

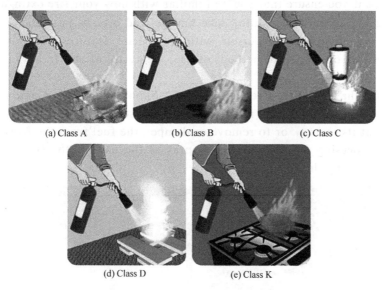

(a) Class A   (b) Class B   (c) Class C

(d) Class D   (e) Class K

Fig. 3. 4 The type of extinguisher

• **Class A**: This is suitable for cloth, wood, rubber, paper, various plastics, and regular combustible fires. It is usually filled with 2 1/2 gallons (9.46 litres) of pressurized water.

• **Class B**: This is suitable for grease, gasoline or oil-based fires. It is usually filled with a dry chemical. Extinguishers smaller than 6lbs (2.72kg) are not recommended.

• **Class C**: This is suitable for electrical fires caused by appliances, tools, and other plugged in gear. It can contain either halon or CO_2. Halon 1211 and 1301 is very expensive and depletes the ozone layer, but it is being replaced by non-depleting agents such as FM200. Note that halon is now illegal in numerous jurisdictions.

• **Class D**: This is used for water-reactive metals such as burning magnesium and will be located in factories using such metals. It comes in the form of a powder that must cover the material to extinguish it.

• **Class K**: This contains a special purpose wet chemical agent for use in kitchen fires and deep fryers to stop fires started by vegetable oils, animal fats, or other fats started in cooking appliances.

• Note that many fire extinguishers will work on a combination of fire classes. You'll need to decide quickly on what type of fire you have and ensure that your fire extinguisher is compatible with the fire you are attempting to extinguish. An all-purpose ABC dry chemical (10lb/4.5kg) extinguisher is a safe bet for most fires, especially where you're not sure of the fire's origins.

5. Ready the fire extinguisher (see Fig. 3.5). Almost all fire extinguishers have a safety pin in the handle. This pin usually looks like a plastic or metal ring, sometimes colored red, that is held in place by a plastic seal. The distinctive features will vary depending on the type of fire extinguisher you have. You must break the seal and pull the safety pin from the handle before you can use the fire extinguisher by squeezing the lever, which discharges the fire extinguishing agent.

• It helps if you ensure that you're familiar with how your fire extinguisher works before being placed under pressure; take time to read over its instructions after you've read this article. Familiarize yourself with its special features and parts. Different extinguishers rely on different methods of use: be aware of this in advance of having to use them.

6. Aim for the base of the fire (see Fig. 3.6). Shooting into the flame is a waste of the fire extinguisher, as you're not putting out the source of the flames. It's vital to stop the fire at its source, or to remove or dampen the fuel from the fire, in order to put it out. By focusing the extinguisher's spray at the base of the fire or the source, you're extinguishing the fuel.

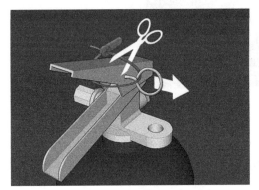
Fig. 3.5 Ready the fire extinguisher

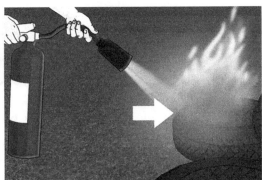
Fig. 3.6 Aim for the base of the fire

7. Remember the simple acronym P. A. S. S. to help you use the fire extinguisher effectively (see Fig. 3.7). P. A. S. S. stands for: Pull, Aim, Squeeze, Sweep, explained below the printable diagram.

• **Pull** the safety pin from the handle. The pin is located at the top of the fire extinguisher. Once removed, it releases the locking mechanism, allowing you to discharge the extinguisher.

• **Aim** the extinguisher nozzle or hose at the base of the fire. As explained, this removes the source or fuel of the fire. Keep yourself low.

• **Squeeze** the handle or lever slowly to discharge the agent. Letting go of the handle will stop the discharge, so keep it held down.

• **Sweep** side to side approximately 6in or 15cm over the fire until expended. The sweeping motion helps to extinguish the fire. Stand several feet or metres back from the fire: fire extinguishers are manufactured for use from a distance.

• The fire may flare up somewhat as extinguishing begins due to the flames being pushed away from the burning material (the real target) by the agent and gust of propellant. Do not be alarmed so long as it dies back promptly.

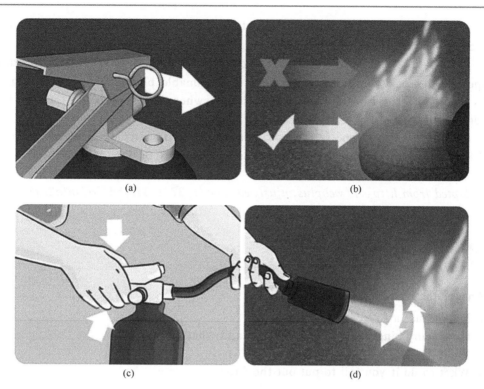

Fig. 3.7 P. A. S. S.

8. Be aware that the typical fire extinguisher will contain around 10 seconds of extinguishing power. If the extinguisher has already been discharged partially, this time will be less.

- If the fire doesn't respond well after you've used up the fire extinguisher, remove yourself to safety quickly.
- If the room fills with smoke, make a hasty exit.

9. Tend to the area if you have successfully put out the fire. This means not leaving it alone, as it might re-ignite without warning. If it is safe to do so, remove fuel sources and commence cleaning up. Water can be used to ensure there are no remaining sparks in materials that are safe to extinguish with water (most combustibles other than oils and other insoluble combustible liquids or places involving electricity). The fire department could help you ensure the fire is completely extinguished. If you're legally obliged to report fires to local authorities, then do so, especially if filing for an insurance claim.

10. Purchase a new fire extinguisher immediately. The old one is now depleted and will serve no further purpose. Do not allow an empty extinguisher to be present where it could create the false impression of being a good extinguisher. A multi-purpose extinguisher is best for a home; check that it is labeled by an independent testing laboratory. Some fire extinguishers can be recharged; for smaller ones, replacing may be cheaper.

- Fire extinguishers should be wall mounted in an accessible place. Keep out of the reach of children who are not responsible enough to leave well alone.

- It's a good idea to always keep a fire extinguisher in the kitchen away from sources of heat such as the stove or cooking surfaces.
- Other good places to keep a fire extinguisher include: your car, your garage (especially if you use welding equipment or flammable products), your caravan or RV, and your boat. In each case, mount it somewhere accessible and protected from outdoor elements.
- Ensure that everyone in the house knows where the fire extinguisher is located and how to use it (provided they are old enough and responsible enough to do so).

Adapted from http://webplus.njust.edu.cn/s/37/t/549/58/be/info22718.html

Comprehension

Answer the following questions.

1. How to use a fire extinguisher?
2. How many kinds of fire extinguishers and what are they?
3. What does P.A.S.S mean?
4. What to do if you fail to put out the fire?

Supplementary knowledge

Structure of Material Safety Data Sheet

1. Chemical Product and Company Identification
PRODUCT NAME: BENZENE (AMOCO/TOTAL)
MANUFACTURER/SUPPLIER: EMERGENCY HEALTH INFORMATION:

Amoco Oil Company
200 East Randolph Drive
Chicago, Illinois 60601 U.S.A.

1 (800) 447-8735
EMERGENCY SPILL INFORMATION:
1 (800) 424-9300 CHEMTREC (USA)
OTHER PRODUCT SAFETY INFORMATION:
(312) 856-3907

2. Composition/information on Ingredients (see Table 3.2)

Table 3.2 Composition on ingredients

Component	CAS#	Range/% by wt.
Benzene	71-43-2	99.80
Toluene	108-88-3	0.20

3. Hazards Identification

EMERGENCY OVERVIEW: Danger! Extremely flammable. Causes eye and skin irritation. Inhalation causes headaches, dizziness, drowsiness, and nausea, and may lead to unconsciousness. Harmful or fatal if liquid is aspirated into lungs. Danger! Contains Benzene. Cancer hazard. Can cause blood disorders. Harmful when absorbed through the skin.

POTENTIAL HEALTH EFFECTS:

EYE CONTACT: Causes mild eye irritation.

SKIN CONTACT: Causes mild skin irritation. Causes skin irritation on prolonged or repeated contact. Harmful when absorbed through the skin.

INHALATION: Cancer hazard. Can cause blood disorders. Inhalation causes headaches, dizziness, drowsiness, and nausea, and may lead to unconsciousness. See "Toxicological Information" section (Section 11.0).

INGESTION: Harmful or fatal if liquid is aspirated into lungs. See "Toxicological Information" section (Section 11.0).

HMIS CODE: (Health: 2)(Flammability: 3)(Reactivity: 0)

NFPA CODE: (Health: 2)(Flammability: 3)(Reactivity: 0)

4. First Aid Measures

EYE: Flush eyes with plenty of water for at least 15 minutes. Get medical attention if irritation persists.

SKIN: Wash exposed skin with soap and water. Remove contaminated clothing, including shoes, and thoroughly clean and dry before reuse. Get medical attention if irritation develops.

INHALATION: If adverse effects occur, remove to uncontaminated area. Give artificial respiration if not breathing. Get immediate medical attention.

INGESTION: If swallowed, drink plenty of water, do NOT induce vomiting. Get immediate medical attention.

5. Fire Fighting Measures

FLASHPOINT: 12°F (-11°C)

UEL: 8.0%

LEL: 1.5%

AUTOIGNITION TEMPERATURE: 928°F (498°C)

FLAMMABILITY CLASSIFICATION: Extremely Flammable Liquid.

EXTINGUISHING MEDIA: Agents approved for Class B hazards (e.g., dry chemical, carbon dioxide, foam, steam) or water fog.

UNUSUAL FIRE AND EXPLOSION HAZARDS: Extremely flammable liquid. Vapor may explode if ignited in enclosed area.

FIRE-FIGHTING EQUIPMENT: Firefighters should wear full bunker gear, including a positive pressure self-contained breathing apparatus.

PRECAUTIONS: Keep away from sources of ignition (e.g., heat and open flames). Keep container closed. Use with adequate ventilation.

HAZARDOUS COMBUSTION PRODUCTS: Incomplete burning can produce carbon monoxide and/or carbon dioxide and other harmful products.

6. Accidental Release Measures

Remove or shut off all sources of ignition. Remove mechanically or contain on an absorbent material such as dry sand or earth. Increase ventilation if possible. Wear respirator and spray with water to disperse vapors. Keep out of sewers and waterways.

7. Handling and Storage

HANDLING: Use with adequate ventilation. Do not breathe vapors. Keep away from ignition sources (e.g., heat, sparks, or open flames). Ground and bond containers when transferring materials. Wash thoroughly after handling. After this container has been emptied, it may contain flammable vapors; observe all warnings and precautions listed for this product.

STORAGE: Store in flammable liquids storage area. Store away from heat, ignition sources, and open flame in accordance with applicable regulations. Keep container closed. Outside storage is recommended.

8. Exposure Controls/Personal Protection

EYE: Do not get in eyes. Wear eye protection.

SKIN: Do not get on skin or clothing. Wear protective clothing and gloves.

INHALATION: Do not breathe mist or vapor. If heated and ventilation is inadequate, use supplied-air respirator approved by NIOSH/MSHA.

ENGINEERING CONTROLS: Control airborne concentrations below the exposure guidelines.

EXPOSURE GUIDELINES: see table 3.3.

Table 3.3 Exposure Limits

Component	CAS#	Exposure Limits
Benzene	71-43-2	OSHA PEL: 1ppm❶ OSHA STEL: 5 ppm ACGIH TLV-TWA: 10ppm
Toluene	108-88-3	OSHA PEL: 100ppm (1989); 200ppm (1971) OSHA STEL: 150ppm (1989); Not established (1971) OSHA Ceiling: 300ppm (1971) ACGIH TLV-TWA: 50ppm (skin)

9. Chemical and Physical Properties

APPEARANCE AND ODOR: Liquid. Colorless. Sweet odor.

pH: Not determined.

VAPOR PRESSURE: 74.6mmHg❷ at 20℃

VAPOR DENSITY: Not determined.

BOILING POINT: 176°F (80℃)

MELTING POINT: 42°F (6℃)

SOLUBILITY IN WATER: Slight, 0.1% to 1.0%.

SPECIFIC GRAVITY (WATER=1): 0.88

10. Stability and Reactivity

STABILITY: Stable.

---

❶ 1ppm=1×10^{-6}，在中国写作 μg/g 或 μL/L。

❷ 1mmHg=133.32Pa。

CONDITIONS TO AVOID: Keep away from ignition sources (e.g. heat, sparks, and open flames).

MATERIALS TO AVOID: Avoid chlorine, fluorine, and other strong oxidizers.

HAZARDOUS DECOMPOSITION: None identified.

HAZARDOUS POLYMERIZATION: Will not occur.

11. Toxicological Information

ACUTE TOXICITY DATA:

EYE IRRITATION: Testing not conducted. See Other Toxicity Data.

SKIN IRRITATION: Testing not conducted. See Other Toxicity Data.

DERMAL LD50: Testing not conducted. See Other Toxicity Data.

ORAL LD50: 3.8g/kg (rat).

INHALATION LC50: 10000ppm (rat).

OTHER TOXICITY DATA: Acute toxicity of benzene results primarily from depression of the central nervous system (CNS). Inhalation of concentrations over 50 ppm can produce headache, lassitude, weariness, dizziness, drowsiness, or excitation. Exposure to very high levels can result in unconsciousness and death.

Long-term overexposure to benzene has been associated with certain types of leukemia in humans. In addition, the International Agency for Research on Cancer (IARC) and OSHA consider benzene to be a human carcinogen. Chronic exposures to benzene at levels of 100 ppm and below have been reported to cause adverse blood effects including anemia. Benzene exposure can occur by inhalation and absorption through the skin.

Inhalation and forced feeding studies of benzene in laboratory animals have produced a carcinogenic response in a variety of organs, including possibly leukemia, other adverse effects on the blood, chromosomal changes and some effects on the immune system. Exposure to benzene at levels up to 300 ppm did not produce birth defects in animal studies; however, exposure to the higher dosage levels (greater than 100 ppm) resulted in a reduction of body weight of the rat pups (fetotoxicity). Changes in the testes have been observed in mice exposed to benzene at 300 ppm, but reproductive performance was not altered in rats exposed to benzene at the same level.

Aspiration of this product into the lungs can cause chemical pneumonia and can be fatal. Aspiration into the lungs can occur while vomiting after ingestion of this product. Do not siphon by mouth.

12. Ecological Information

Ecological testing has not been conducted on this product.

13. Disposal Information

Disposal must be in accordance with applicable federal, state, or local regulations. Enclosed-controlled incineration is recommended unless directed otherwise by applicable ordinances. Residues and spilled material are hazardous waste due to ignitability.

14. Transportation Information

U.S. DEPT OF TRANSPORTATION

Shipping Name　　　　　　　　Benzene

Hazard Class	3
Identification Number	UN1114
Packing Group	II
RQ	RQ

INTERNATIONAL INFORMATION:
　　Sea (IMO/IMDG)
　　　　Shipping Name Not determined.
　　Air (ICAO/IATA)
　　　　Shipping Name Not determined.
　　European Road/Rail (ADR/RID)
　　　　Shipping Name Not determined.
　　Canadian Transportation of Dangerous Goods
　　　　Shipping Name Not determined.

15. Regulatory Information

CERCLA SECTIONS 102a/103 HAZARDOUS SUBSTANCES (40 CFR Part 302.4): This product is reportable under 40 CFR Part 302.4 because it contains the following substance (s) (see table 3.4).

Table 3.4　Component information

Component/CAS Number	Weight/%	Component Reportable Quantity (RQ)
Benzene 71-43-2	99.80	10lbs. ❶

　　SARA TITLE III SECTION 302 EXTREMELY HAZARDOUS SUBSTANCES (40 CFR Part 355): This product is not regulated under Section 302 of SARA and 40 CFR Part 355.

　　SARA TITLE III SECTIONS 311/312 HAZARDOUS CATEGORIZATION (40 CFR Part 370): This product is defined as hazardous by OSHA under 29 CFR Part 1910.1200 (d).

　　SARA TITLE III SECTION 313 (40 CFR Part 372): This product contains the following substance (s), which is on the Toxic Chemicals List in 40 CFR Part 372 (see table 3.5).

Table 3.5　Component information

Component/CAS Number	Weight Percent/%
Benzene 71-43-2	99.80

　　U.S. INVENTORY (TSCA): Listed on inventory.
　　OSHA HAZARD COMMUNICATION STANDARD: Flammable liquid. Carcinogen. Irritant. CNS Effects. Target organ effects.
　　EC INVENTORY (EINECS/ELINCS): In compliance.
　　JAPAN INVENTORY (MITI): Not determined.
　　AUSTRALIA INVENTORY (AICS): Not determined.
　　KOREA INVENTORY (ECL): Not determined.
　　CANADA INVENTORY (DSL): Not determined.
　　PHILIPPINE INVENTORY (PICCS): Not determined.

　　　　Adapted from http：//hazard.com/msds/mf/amoco/files/11697000.html

---

❶　lbs，磅的复数形式，1lb＝0.4536kg。

Product Testing

Objectives:

After finishing this project, you are able to:
- Name analytical instruments properly
- Do a Redox Titration and related calculation
- Select appropriate acid-base indicators according to the characteristics of the reactions

Warming-up

What Are the Usages of the Following Instruments?

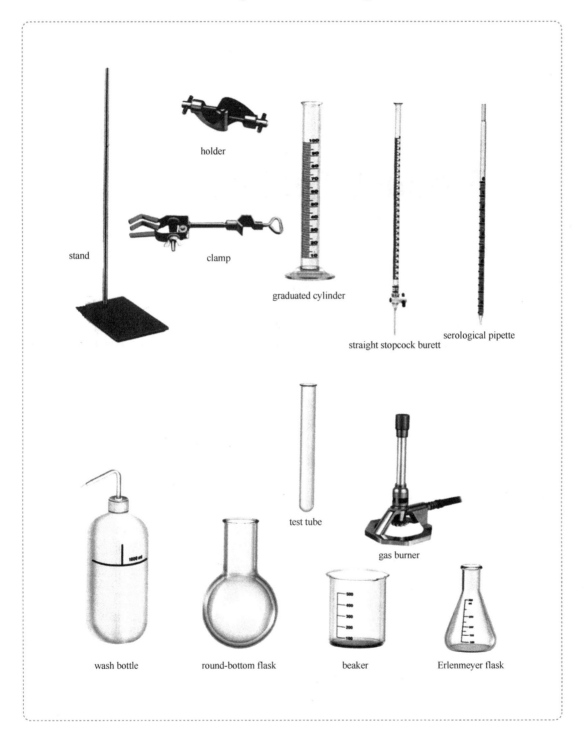

Text A Instruments

Review the steps of using a buret and a pipet. Speculate the meaning of the words with the underlines

Words and expressions

Buret

A buret is used to deliver solution in precisely measured, variable volumes. Burets are used primarily for titration, to deliver one reactant until the precise end point of the reaction is reached.

buret [bjuˈrɛt] n. [分化] 滴定管
solution [səˈluʃən] n. 溶液

titration n. [分化] 滴定

Using a Buret

To fill a buret, close the stopcock at the bottom and use a funnel. You may need to lift up on the funnel slightly, to allow the solution to flow in freely.

stopcock [ˈstɑpkɑk] n. 活塞

funnel [ˈfʌnl] n. 漏斗

You can also fill a buret using a disposable transfer pipet. This works better than a funnel for the small, 10mL burets. Be sure the transfer pipet is dry or conditioned with the titrant, so the concentration of solution will not be changed.

transfer pipet 移液管

titrant [ˈtaɪtrənt] n. 滴定液
concentration [ˌkɑnsnˈtreʃən] n. 浓度

Before titrating, condition the buret with titrant solution and check that the buret is flowing freely. To condition a piece of glassware, rinse it so that all surfaces are coated with solution, then drain. Conditioning two or three times will insure that the concentration of titrant is not changed by a stray drop of water.

rinse [rɪns] vt. 冲洗

drain [dren] vi. 排水

stray [stre] adj. 散落的

Check the tip of the buret for an air bubble. To remove an air bubble, whack the side of the buret tip while solution is flowing. If an air bubble is present during a titration, volume readings may be in error. Rinse the tip of the buret with water from a wash bottle and dry it carefully. After a minute, check for solution on the tip to see if your buret is leaking. The tip should be clean and dry before you take an initial volume reading.

initial [ɪˈnɪʃəl] adj. 最初的

When your buret is conditioned and filled, with no air bubbles or leaks, take an initial volume reading. A buret reading card with a black rectangle can help you to take a more accurate reading. Read the bottom of the meniscus. Be sure your eye is at the level of meniscus, not above or below. Reading from an angle, rather than straight on, results in a

rectangle [ˈrɛktæŋgl] n. 矩形

meniscus [məˈnɪskəs] n. 弯液面

parallax error.

Deliver solution to the titration flask by turning the by turning the stopcock. The solution should be delivered quickly until a couple of mL from the endpoint.

The endpoint should be approached slowly, a drop at a time. Use a wash bottle to rinse the tip of the buret and the sides of the flask.

Pipet

A pipet is used to measure small amounts of solution very accurately. A pipet bulb is used to draw solution into the pipet.

Using a Pipet

Start by squeezing the bulb in your preferred hand. Then place the bulb on the flat end of the pipet. Place the tip of the pipet in the solution and release your grip on the bulb to pull solution into the pipet. Draw solution in above the mark on the neck of the pipet. If the volume of the pipet is larger than the volume of the pipet bulb, you may need to remove the bulb from the pipet and squeeze it and replace it on the pipet a second time, to fill the pipet volume completely.

Quickly, remove the pipet bulb and put your index finger on the end of the pipet. Gently release the pipet. Gently release the seal made by your finger until the level of the solution meniscus exactly lines up with the mark on the pipet. Practice this with water until you are able to use the pipet and bulb consistently and accurately.

parallax ['pærəlæks] n. 视差

grip [grɪp] n. 紧握

seal [siːl] n. 密封

consistently [kən'sɪstəntli] adv. 一致地

Adapted from "schoeff, M. S. and Willianms, R. H. Principles of Laboratory Instruments, Mosby: St. Louis, 1993"

Comprehension

Choose the best answer according to the text.

1. In titration operation, the titrant solution is volumetrically delivered to the reaction flask using a (　　).
   A. buret　　　　　　　　B. Balance
   C. pipet　　　　　　　　D. tube

2. When you take the volume reading in a titration operation, your eyes should be (　　) the level of meniscus.
   A. above　　　　　　　　B. below

C. at D. near

3. Rinse the tip of the buret with water from a (　　) and dry it carefully.
   A. pipet B. flask
   C. bulb D. wash bottle
4. Which order is right for titration operation? (　　)
   A. condition, take an volume reading, check for bubble, calculation
   B. condition, check for bubble, take an volume reading, calculation
   C. check for bubble, condition, take an volume reading, calculation
   D. check for bubble, condition, calculation, take an volume reading

Vocabulary building

Active words

solution
n. 解决方案；溶液

concentration
n. 浓度；集中，集合

present
adj. 现在的；出席的
n. 现在；礼物

vt. 介绍；呈现

approach
n. 方法
v. 接近

draw
vt. 画；吸引

Useful expressions

rather than 而不是
result in 导致

check for 检查
line up with 与……对齐

Exercise

Fill in each blank with a given word or expression in their right form.

result in　　rather than　　approach　　present　　concentration　　solution

1. There is a _____ of people in big cities
2. $KMnO_4$ will react with the small amount organic compound _____ in the water when first prepared.
3. The gas released by motor vehicle _____ air pollution.
4. We need more joined-up thinking in our _____ to the environment.
5. Water precipitates camphor（樟脑）from its alcoholic _____.
6. I, _____ you, should do the work.

Extension

Common analytical apparatus

离心机	centrifuge	电炉	heater
磁力搅拌器	magnetic stirrer	闪点仪	flash point tester
折光仪	refractometer	容量瓶	measuring flask
pH 计	pH meter	烘箱	oven
pH 试纸	pH indicator paper	蒸发皿	evaporating dish
蒸馏装置	distilling apparatus	表面皿	watch glass
酒精灯	alcohol burner	冷凝器	condenser
滤管	filter	滴定管	burette
移液管	pipette	空气过滤器	air filter

Abbreviation of analytical methods

CEP	capillary electrophoresis	毛细管电泳法
GC	gas chromatography	气相色谱法
HPLC	high performance liquid chromatography	高效液相色谱法
IC	ion chromatography	离子色谱法
IR	infrared spectroscopy	红外光谱法
LC	liquid chromatography	液相色谱法
MS	mass spectrometry	质谱法
NMR	nuclear magnetic resonance spectrometry	核磁共振波谱法
PC	paper chromatography	纸色谱法
PE	paper electrophoresis	纸电泳法
RPLC	reversed phase liquid chromatography	反相液相色谱法
SP	spectrophotometry	分光光度法
TLC	thin-layer chromatography	薄层色谱法
UVS	ultraviolet spectrometry	紫外光谱法
UVF	ultraviolet fluorescence spectrometry	紫外荧光光谱法

Text B Redox Titration

Introductory Theory

Oxidation is the name given to a chemical process in which an atom or ion loses one or more electrons. Reduction is the name given to chemical process in which an atom or ion gains one or more electrons. Obviously, in order for a species to gain an electron, some other species has to lose an electron. Therefore, oxidation and reduction always go together. In fact, the two combined processes are often known as "redox".

Substances, which are reduced, do so by causing other objects to be oxidized, so they are often referred to as oxidizing agents. Substances, which are oxidized, do so by causing other objects to be reduced, so they are known as reducing agents.

In this experiment, a redox reaction will be performed in order to help us do a quantitative analysis. We have a solution of $KMnO_4$, a strong oxidizing agent. We know that its concentration is approximately 0.01 mol/L, but we cannot prepare a solution of it whose concentration is accurately known. This is because the $KMnO_4$ will react with small amounts of organic material present in the water when the solution is first prepared. As a result, the concentration will never be as high as calculated. We are going to react a known volume of $KMnO_4$ with a known volume of a reducing agent $FeSO_4(NH_4)_2SO_4 \cdot 6H_2O$. The concentration of this solution can be accurately known. We will perform stoichiometric calculations to determine the concentration of the potassium permanganate. The combination of the two solutions gives the net ionic equation:

$$5Fe^{2+} + MnO_4^- + 8H^+ \longrightarrow Mn^{2+} + 5Fe^{3+} + 4H_2O$$

Unlike acid-base titration, where an indicator is required, this reaction is self-indicating. The permanganate ions give their solution a very dark purple color. As they react with the ferrous ions, the purple disappears. If even a fraction of a drop of permanganate is added after the equivalence point (the point when just enough permanganate has been added to react with all of the iron present), there will be no ferrous ions to remove the color, and there will be a persistent pink color. It is critical to titrate only until the faintest persistent pink color can be observed.

Procedure

(1) 100mL of 0.0500mol/L Fe^{2+} solution was prepared using 40mL of 3mol/L H_2SO_4 and 1.96g of the iron (II) ammonium sulfate hexahydrate.

(2) The titration apparatus was prepared and the burette was filled with $KMnO_4$ of unknown concentration. 10mL of the 0.0500mol/L Fe^{2+} solution was put into a flask.

(3) Three titrations of the unknown $KMnO_4$ solution into 10mL of the Fe^{2+} solution were performed and the data was recorded. The end point occurs when a very faint pink color persist.

(4) Watch to make sure that you do not go over the end point. Even one drop too much will make the solution a very dramatic pink color (see Table 4.1).

Table 4.1 Experiment data

Project	Trial 1	Trial 2	Trial 3
Volume of Fe^{2+} solution/mL	10.0	10.0	10.0
Initial burette reading/mL	15.82	31.08	49.39
Final burette reading/mL	5.05	21.52	38.10
Volume of $KMnO_4$/mL	10.77	9.56	10.29
Concentration of $KMnO_4$/mol/L	0.009285	0.010500	0.009718

Data Table

Analysis-sample Calculations:

$5Fe^{2+} + MnO_4^- + 8H^+ \longrightarrow Mn^{2+} + 5Fe^{3+} + 4H_2O$

 0.0500mol/L x mol/L

 10.0mL   10.77mL

0.0500 mol/L Fe^{2+} = (0.0500mol Fe^{2+}/L) × 0.0100L × (1mol MnO_4^-/5mol Fe^{2+}) × (1/0.01077L) = 0.00929mol/L MnO_4^-

Average $KMnO_4$ concentrations: 0.00984mol/L

Conclusion

The average molar concentration of the aqueous potassium permanganate solutions used in the lab's three trial was 0.00984 mol/L.

ferrous ['fɛrəs] adj. [化学] 亚铁的
equivalence point 等当点, 滴定终点
persistent [pə'zɪstənt] adj. 坚持的

trial ['traɪəl] n. 试验

aqueous ['ekwɪəs] adj. 水的

Adapted from "Haugland, R. P., Handbook of Fluorescnet Probes and Research Chemicals; Molecular Probes Inc, ; Eugene, OR, 1985."

Comprehension

Choose the best answer according to the text.

1. Ox + Red ⟶ Red´ + Ox´. This reaction is of (　　) type.
   A. complex formation　　　　　B. Acid-base
   C. precipitation　　　　　　　　D. oxidation-reduction

2. (　　) is the name given to a chemical process in which an atom or ion gains one or more electrons.
   A. Oxidation　　　　　　　　　B. Reduction
   C. Titration　　　　　　　　　　D. Analysis

3. Usually, (　　) titrations are done and the data was recorded.
   A. two　　　B. three　　　C. four　　　D. five

4. Why we cannot prepare a solution of $KMnO_4$ whose concentration is accurately known?
   A. Because the apparatus are not so accurate
   B. Because there are reactive materials present in the water
   C. Because $KMnO_4$ is a strong oxidizing agent
   D. Because $KMnO_4$ solution is not first prepared

5. Titrate the unknown $KMnO_4$ solution into Fe^{2+} solution, and the solution will be of (　　) color at the end point.
   A. dark purple　　　　　　　　B. transparent
   C. faintest persistent pink　　　D. dramatic pink

Vocabulary building

Active words

reduce
v. 减少，降低；还原

quantitative
adj. 定量的

analysis
n. 分析

process
n. 过程；工艺
vt. 加工，处理

perform
v. 表演；执行，进行

agent
n. 代理人，代理商；药剂
vt. 由……作中介；由……代理

organic
adj. [化] 有机的

fraction
n. 分数；部分；馏分

Useful expressions

be referred to as　被称为……
small amounts of　少量的
give…color　使某物呈现出某色
make sure that　确保

Exercise

Fill in each blank with a given word or expression in their right form.

quantitative   reduce   analysis   process   perform   organic

1. The reaction is generally _____ in a flask containing the liquid or dissolved sample.

2. This company is strong in chemical and petrochemical, automobile and engineering, agro and food _____.

3. We know the constitution of mineral through chemical _____.

4. Today we will express this more _____ in three laws which are called Newton's Laws.

5. BPA is an important _____ chemical material, is mainly used to produce various high molecular materials.

6. The side reaction was _____, so the current efficiency of electrolyzer was raised.

Work in groups

Discuss with your group members about translation method of decimals, fractions and multiples.

12.89	Twelve point eight nine	数十/几十	Tens of / Dozens of
6.834	Six point eight three four	几百	Hundreds of
0.256	Point two five six	几万	Tens and thousands of
0.654	Zero point six five four	成千上万	Millions of
2/3	Two thirds	二十有余	Twenty [and] odd / More than twenty / Over twenty / Above twenty
450	Four hundred and fifty		
23600	Twenty-three thousand and six hundred		
36%	Thirty-three percent	四百多	Four hudred [or] odd
10^6	Ten to the power of six / Ten to the sixth power / Ten to the sixth / Ten to the six	30 克左右	Thirty grams or so / About thirty grams / Some thirty grams
		≤3.6%	Up to three point six perctn
		≥3.5g/L	Down to three point five grams per liter
10^{-3}	Ten to the power of minus three	减少 70%	Decrease 70% / Reduce by 70% / 70% less
3.4×10^6	Three point four times ten to the negative six	减少到 70	Decrease to 70
		减少 1/3	Decrease three times

Translate the following numbers into English.

1. 用水将溶液稀释到原体积的3倍。

_____

2. 使用新型催化剂后，反应速率提高了60％。

_____

3. 因为高锰酸钾是强氧化剂，所以新配制出的溶液浓度应小于计算值。

Discuss with your group members about common suffix of the numbers.

词冠	符号	10 的因数	词冠	符号	10 的因数
kilo	k	10^{3}	milli	m	10^{-3}
hecto	h	10^{2}	micro	μ	10^{-6}
deka (deca)	da	10^{1}	nano	n	10^{-9}
deci	d	10^{-1}	pico	p	10^{-12}
centi	c	10^{-2}			

Text C Acid-Base Indicators

Words and expressions

Acid-Base indicators (also known as pH indicators) are substances which change color with pH. They are usually weak acids or bases, which when dissolved in water dissociate slightly and form ions.

Consider an indicator which is a weak acid, with the formula HIn. At equilibrium, the following equilibrium equation is established with its conjugate base:

$$HIn(aq) + H_2O(l) \rightleftharpoons H_3O^+(aq) + In^-(aq)$$
  Acid   conjugate base
 (color A)   (color B)

The acid and its conjugate base have different colors. At low pH values the concentration of H_3O^+ is high and so the equilibrium position lies to the left. The equilibrium solution has the color A. At high pH values, the concentration of H_3O^+ is low—the equilibrium position thus lies to the right and the equilibrium solution has color B.

Phenolphthalein is an example of an indicator which establishes this type of equilibrium in aqueous solution:

indicator ['ɪndɪkeɪtə] n. 指示剂
dissolve [dɪ'zɒlv] v. 溶解
dissociate [dɪ'səʊʃɪeɪt] vi. 电离
formula ['fɔːmjələ] n. 公式
equilibrium [ˌiːkwɪ'lɪbrɪəm] n. 平衡
equation [ɪ'kweɪʒn] n. 方程式
conjugate ['kɒndʒəget] adj. 共轭的

position [pə'zɪʃən] n. 位置

phenolphthalein [ˌfiːnɒl'(f)θeɪliːn] n. [试剂] 酚酞
aqueous ['eɪkwɪəs] adj. 水的

colorless(Acid) + H_2O ⇌ pink(Base) + H_3O^+

Phenolphthalein is a colorless, weak acid which dissociates in water forming pink anions. Under acidic conditions, the equilibrium is to the left, and the concentration of the anions is too low for the pink color to be observed. However, under alkaline conditions, the equilibrium is to the right, and the concentration of the anion becomes sufficient for the pink color to be observed.

We can apply equilibrium law to indicator equilibria in general for a weak acid indicator:

$$K_{In} = \left(\frac{[H_3O^+][In^-]}{[HIn]}\right)_{eq}$$

K_{In} is known as the indicator dissociation constant. The color of the indicator turns from color A to color B or vice versa at its turning point. At this point:

$$[HIn] = [In^-]$$

So from equation:

$$K_{In} = [H_3O^+]$$

The pH of the solution at its turning point is called the pK_{In} and is the pH at which half of the indicator is in its acid form and the other half in the form of its conjugate base.

At a low pH, a weak acid indicator is almost entirely in the HIn form, the color of which predominates. As the pH increases, the intensity of the color of HIn decreases and the equilibrium is pushed to the right. Therefore the intensity of the color of In^- increases. An indicator is most effective if the color change is distinct and over a low pH range. For most indicators the range is within ±1 of the pK_{In} value.

A Universal Indicator is a mixture of indicators which give a gradual change in color over a wide pH range—the pH of a solution can be approximately identified when a few drops of universal indicator are mixed with the solution. Indicators are used in titration solutions to signal the completion of the acid-base reaction.

anion ['ænaɪən] n. 阴离子

alkaline ['ælkə'laɪn] adj. 碱性的

constant ['kɒnstənt] n. [数] 常数
vice versa [ˌvaisi'və:sə] 反之亦然

predominate [prɪ'dɒmə'net] vi. 占主导（或支配）地位；占优势
intensity [ɪn'tənsəti] n. 强度

distinct [dɪ'stɪŋkt] adj. 有区别的

universal [ˈjʊnə'vɜsl] adj. 通用的

identify [aɪ'dɛntɪfaɪ] v. 确定

signal ['sɪgnl] vt. 标志

Adapted from "Wellinder, B. S.; Kornfelt, T.; Sorensen, H. H.; Analytical Chemistry, 1995, Vlo. 67, pp.39-43"

Comprehension

Choose the best answer according to the text.

1. Acid-base indicators are substances which change colors with (　　).
   A. temperature　　　B. pressure　　　C. pH　　　D. concentration
2. Phenolphthalein is an example of a/an (　　) indicator.
   A. acidic　　　B. basic　　　C. neutral
3. Consider an indicator, which is a weak acid, with the formula HIn. At low pH, its aqueous solution will assumes the color of (　　).
   A. HIn　　　B. In^-　　　C. H^+　　　D. H_3O^+
4. A/An (　　) indicator is a mixture of indicators, which give a gradual change in color over a wide pH range.
   A. universal　　　B. acidic　　　C. basic　　　D. neutral

Vocabulary building

Active words

acid
n. 酸

base
n. 基础；碱

indicator
n. 指示器；[试剂] 指示剂；[计] 指示符；压力计

aqueous
adj. 水的

apply
v. 申请；应用

constant
adj. 经常的，不变的
n. 常数，恒量

equilibrium
n. 均衡；平衡

equation
n. 方程式，等式

Useful expressions

vice versa 反之亦然
lie to 位于

under…condition 在……条件下

Exercise

Match the word or phrase with their equivalent.

1. equilibrium　　A. similar to or containing or dissolved in water
2. equation　　　B. a quantity that does not vary
3. aqueous　　　C. with the order reversed
4. acid　　　　　D. a chemical reaction and its reverse proceed at equal rates
5. constant　　　E. a mathematical statement that two expressions are equal
6. vice versa　　F. any of various water-soluble compounds having a sour taste and capable of turning litmus red and reacting with a base to form a salt

Extension

Indicator Range

Indicators	pH range
Methyl violet	0.0-1.6
Malachite green	0.2-1.8
Thymol blue	1.2-2.8
Methyl yellow	2.9-4.0
Bromphenol blue	3.0-4.6
Congo red	3.0-5.2
Methyl orange	3.1-4.4
Bromcresol green	3.8-5.4
Methyl red	4.2-6.3
litmus	4.5-8.3
Bromcresol purple	5.2-6.8
Bromcresol blue	6.0-7.6
Phenol red	6.6-8.0
Thymol blue	8.0-9.6
Phenolphthalein	8.2-10.0

Reading material

Gas and Liquid Chromatography

In analytical chemistry, gas chromatography is a technique for separating chemical substances in which the sample is carried by a moving gas stream through a tube packed with a finely divided solid that may be coated with a film of a liquid. Because of its simplicity, sensitivity and effectiveness in separating components of mixtures, gas chromatography is one of the most important tools in chemistry. It is widely used for quantitative and qualitative analysis of mixtures. for the purification of compounds, and for the determination of such thermochemical constants as heats of solution and vaporization, vapor pressure, and activity coefficients. Gas chromatography is also used to monitor industrial processes automatically gas streams are analyzed periodically, and manual or automatic responses are made to counteract undesirable variations. Many routine analyses are performed rapidly in medical and other fields. For example, by the use of only 0.1 milliliter of blood, it is possible to determine the percentages of dissolved oxygen, nitrogen, carbon dioxide, and carbon monoxide. Gas chromatography is also useful in the analysis of air pollutants, alcohol in blood essential oils, and food products.

The method consists of, first, introducing the test mixture or sample into a stream of an inert gas, commonly helium or argon, that acts as carrier. Liquid samples are vaporized before injection into the carrier stream. The gas stream is passed through the packed column, through which the components of the sample move at velocities that are influenced by the degree of interaction of each constituent with the stationary nonvolatile phase. The substances having the greater interaction with the stationary phase are retarded to a greater extent and consequently separate from those with smaller interaction. As each component leaves the column with the carrier, it passes through a detector and then either goes to a fraction collector or is discarded.

For liquid chromatography, the procedure can be performed either in a column or on a plane. Columnar liquid chromatography is used for qualitative and quantitative analysis in a manner similar to the way in which gas chromatography is employed. Sometimes retention volumes, rather than retention times, are used for qualitative analysis. For chemical analysis, the most popular category of columnar liquid chromatography is high-performance liquid chromatography (HPLC). The method uses a pump to force one or more mobile phase solvents through high-efficiency, tightly packed columns. As with gas chromatography, an injection system is used to insert the sample into the entrance to the column, and a detector at the end of the column monitors the separated analyte components.

The stationary phase that is used for plane chromatography is physically held in place in or on a plane. Typically the stationary phase is attached to a plastic, metallic, or glass plate. Occasionally, a sheet of high-quality filter paper is used as the stationary phase. The sample is added as a spot or a thin strip at one end of the plane. The mobile phase flows over the spot by capillary action during ascending development or as a result of the force of gravity during descending development. During ascending development, the end of the plane near and below the sample spot is dipped into the mobile phase, and the mobile phase moves up and through the spot. During descending development, the mobile phase is added to the top of the plane and flows downward through the spot.

Qualitative analysis is performed by comparing the retardation factor of the analyte components with the retardation factors of known substances. The retardation factor is defined as the distance from the original sample spot that the component has moved divided by the distance that the mobile phase front has moved and is constant for a solute in a given solvent. Quantitative analysis is performed by measuring the sizes of the developed spots, by measuring some physical property of the spots (such as fluorescence), or by removing the spots from the plane and assaying them by another procedure.

Comprehension

Answer the following questions.

1. Which factors are important for quantitative analysis with chromatography?
2. What is difference between gas chromatography and liquid one?
3. What is the uses of chromatography?
4. What is the definition of retardation factor?

Supplementary knowledge

Structure of Operating Manual

Operating Manual Of Model 722 Visible Spectrophotometer

Chief Uses

Model 722 visible spectrophotometer, an analytical instrument commonly used in physico-chemical laboratories to make quantitative and qualitative analysis of specimen materials in the near ultra-violet, visible spectral range, finds much scope for its service in such fields as medicine, clinical examination, biochemistry, petrochemical industry, environmental protection and quality control.

Working Conditions

① The instrument should be installed in a dry room with the temperature kept between 5℃ and 35℃, and relative humidity under 85%.

② While in use, the instrument should be placed on a solid, smooth working table, free from shocks or vibrations.

③ Strong illumination and direct exposure to sunlight should be avoided.

④ No airflow, e.g., produced by an electric fan, should be directed straight towards the instrument.

⑤ Keep the instrument far away from any high-intensity magnetic field, electric field or electrical equipment that produces high-frequency waves.

⑥ Power supply for the instrument should be AC 220V ± 22V, 50Hz ± 1Hz, with good ground connection. To increase the anti-interference performance, it is recommended to use AC regulated power supply, either an electronic AC voltage stabilizer or an AC constant voltage regulator with a capacity of 1000W or above.

⑦ Do not use the instrument in a place where there is hydrogen sulfide, fluorine sulfite or other corrosive gases.

Main Specifications (see table 4.2)

Table 4.2 Specifications

Names	Specifications
Optical System	Single beam, 1200lines/mm ruled grating
Spectral Bandwidth	6nm
Wavelength Range	325-1000nm
Light Source	Tungsten Halogen Lamp 6V/10W
Wavelength Accuracy	±2.0nm
Wavelength Reproducibility	1.0nm
Wavelength Readout	4 place LED digital display
Stray Light	≤0.5% (T) at 360nm
Photometric Range	0-125%T, -0.097-2.000A, 0—1999C (0—1999℃)
Photometric Accuracy	±1.0%T
Data Output	RS-232 Serial Port
Power Requirement	AC110-220V, 50Hz±1Hz
Printer Output	Serial Port
Dimensions	430mm×310mm×200mm
Weight	8kg

Operation Principles

The 722 spectrophotometer works on the colorimetric principle/Beer's law. The substance in the solution, excited by the light irradiation, produces an effect of light absorption. Each kind of substance has its own absorption spectrum distinct from others because of its absorption selectivity. When a monochromatic light is transmitted through the solution, the light energy decreases as it is absorbed, in proportion to the concentration of the substance (see Fig. 4.1).

$$T = I/I_0$$
$$\lg I_0/I = KcL$$
$$A = KcL$$

In the above: T——Transmittance
I——Intensity of the transmission light
I_0——Intensity of the incident light
A——Absorbance
K——Absorption coefficient
L——Length of light-travel through the solution
B——Concentration of the solution

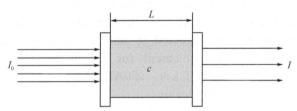

Fig. 4.1 722 Spectrophotometer

It can be inferred from the formula that when the intensity of the incident light, absorption coefficient and length of light travel through the solution remain invariant, transmission light varies according to the concentration of the solution.

Optical Design and Outward Appearance (see Fig. 4.2)

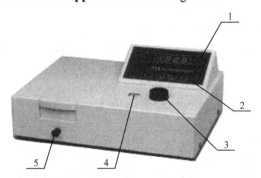

Fig. 4.2 The outward Appearance
1—result display; 2—function key; 3—wavelength drive; 4—wavelength display; 5—push rod of cells holder

The optical system of auto-collimating dispersion grating and single beam light path is adopted for 7205 visible spectrometer. The optical diagram is shown in Fig. 4.3.

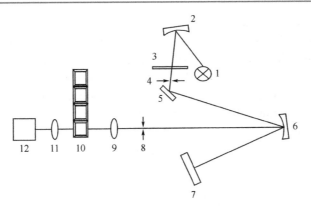

Fig. 4.3 The optical diagram
1—tungsten-halogen lamp; 2—collecting mirror; 3—filter; 4—entrance slit; 5—reflective mirror;
6—collimator; 7—grating; 8—exit slit; 9—condensing lens;
10—sample holder; 11—shutter; 12—photocell

Model 722 visible Spectrophotometer outward appearance is shown below.

Installation, Operation and Maintenance

Before installation, check the safety of the instrument, the voltage of the power supply and the ground connection. Calibrate the instrument before it is put to use, as the wavelength accuracy may be affected by handing in transportation. Turn on the power switch and preheat the instrument for 30 minutes. The four keys on the keyboard are ①ENTER key, ②0%T key, ③0ABS/100%T key, ④FUNCTION key.

① The ENTER key has two functions:

a. To be used in RS232 serial interface and data transmission (one-way transmission from the instrument to the computer).

b. When in F mode, to confirm the current F value, and to switch automatically to C, whose value will be calculated ($c = FA$).

② The 0%T key is used:

a. To set zero, it is effective only in T mode, Uncover the sample compartment, press the key, and it will show 000.0.

b. As the descending key. When in F mode, press the key and the F value will decrease by 1. Holding the key will speed up the decrease. When the F value reaches 0, a press on the key will change the value to 1999.

③ The 0ABS/100%T key has three functions:

a. When in A or T mode, close the sample compartment, press the key and it will read 0.000 or 100.0.

b. To be used as the ascending key (effective only in F mode). Press the key and the F valued will increase by 1. A long press will speed up the increase.

c. When the F valued reaches 1999, another push of the key will change it to 0. And still another push it will increase by 1.

④ The A/T/C/F key is used to switch between A, T, C and F modes.

A—Absorbance;
T—Transmittance;
C—Concentration;
F—Factor.

Troubleshooting (see table 4.2) and Calibration

Table 4.2 Troubleshooting

Troubles	Causes	Remedies
No functioning after power is on	1. Power supply not connected 2. Power fuse melted 3. Loose contact of power switch	1. Check power line 2. Replace fuse 3. Replace power switch
Unstable display	1. Insufficient warm-up time 2. Excessive vibration, strong air current near light source, or strong external light 3. Unstable voltage 4. Poor grounding	1. Ensure 30 minutes warm-up time 2. Improve working conditions 3. Check power voltage 4. Check ground connection
Unable to set zero	Amplifier failure	Repair amplifier
Unable to set 100%	1. Tungsten-halogen lamp not lit 2. Inaccurate light path 3. Amplifier failure	1. Check or repair lamp power circuit 2. Readjust light path 3. Repair amplifier
Concentration Incorrectly calculated	Display board out of order	Repair or replace display board

After a long time of use, the instrument needs calibrating or repairing, as its performance index may have changed. Here are some tips.

① Replacement of tungsten-halogen lamp

put on a clean pair of working gloves before you replace the lamp. Be sure the power is off. Loosen the two clamping screws on the lamp bracket with a wrench. Remove the damaged lamp and fix the new one. Set the wavelength of the instrument at 500nm. Switch on the power, move the lamp up and down and from side to side, until its focus falls exactly on the entrance slit. Observe the reading in T mode without adjusting the 100% T key, readjust the lamp to make the reading at its maximum. Finally, tighten the two screws.

Caution: The two clamping screws serve as the voltage output end of the tungsten-halogen lamp. Be sure that short circuit never occurs when the lamp is turned on.

② Calibration of wavelength accuracy

Spectrophotometer 722 is checked for its wavelength accuracy by a point-to-point method against the two characteristic absorption peaks of a didymium filter—529nm and 808nm. In case the point-to-point measurement shows a wavelength different from the peak wavelength of the didymium filter (maximum permissible error is ±2nm), remove the wavelength knob, loosen the three positioning screws on the wavelength dial, adjust the dial pointer to the characteristic absorption peak wavelength value, and tighten the screws.

Standard Package
① 1 main set;
② 1 power supply cable;
③ 4 glass cells;
④ 1 black body;

⑤ 1　dust cover;
⑥ 1　operating manual;
⑦ 1　packing list;
⑧ 1　certificate of inspection.

Storage and Warranty Period

The instrument should be stored in its original manufacturer's packaging in a room where the ambient temperature is kept between 5℃ and 35℃, relative humidity no more than 85%, and the air contains no harmful, corrosive substance.

The manufacturer will repair the produce free of charge if, during the period of twelve months from the time of its original purchase, this produce proves defective under normal handing, use and service due to defective materials or workmanship (easily worn parts excluded).

Environment Protection

Objectives:

After finishing this module, you are able to:
- Classify different kinds of pollutions
- Handle household hazardous waste properly
- Do a BOD test correctly with a kit

Warming-up

Types of Pollution

Air pollution

Air pollution is termed as the release of chemicals and particulates into the atmosphere. Common gaseous pollutants include carbon monoxide, sulfur dioxide, chlorofluorocarbons (CFCs) and nitrogen oxides produced by industry and motor vehicles.

Noise pollution

Noise pollution encompasses roadway noise, aircraft noise, industrial noise as well as high-intensity sonar.

Thermal pollution

Thermal pollution, is a temperature change in natural water bodies caused by human influence, such as use of water as coolant in a power plant.

Radioactive contamination

Radioactive contamination results from 20th century activities in atomic physics, such as nuclear power generation and nuclear weapons research, manufacture and deployment. (See alpha emitters and actinides in the environment.)

Light pollution

Light pollution includes light trespass, over-illumination and astronomical interference.

Water pollution

discharge of wastewater from commercial and industrial waste into surface waters; discharges of untreated domestic sewage, and chemical contaminants from treated sewage; release of waste and contaminants into surface runoff flowing to surface waters; waste disposal and leaching into groundwater; eutrophication and littering.

Visual pollution

Visual pollution refers to the presence of overhead power lines, motorway billboards, scarred landforms (as from strip mining), open storage of trash, municipal solid waste or space debris.

Littering

Littering refers to the criminal throwing of inappropriate man-made objects, unremoved, onto public and private properties.

Text A Water Pollution

Comprising over 70% of the Earth's surface, water is undoubtedly the most precious natural resource that exists on our planet. Without water, life on the earth would be non-existent: it is essential for everything on our planet to grow and prosper. Although we as humans recognize this fact, we disregard it by polluting our rivers, lakes, and oceans. Subsequently, we are slowly but surely harming our planet to the point where organisms are dying at a very alarming rate. In addition to innocent organisms dying off, our drinking water has become greatly affected as is our ability to use water for recreational purposes. In order to combat water pollution, we must understand the problems and become part of the solution.

Causes of Pollution

Many causes of pollution including sewage and fertilizers contain nutrients such as nitrates and phosphates. In excess levels, nutrients over stimulate the growth of aquatic plants and algae. Excessive growth of these types of organisms consequently clogs our waterways, uses up dissolved oxygen as they decompose, and blocks light to deeper waters. This in turn proves very harmful to aquatic organisms as it affects the respiration ability of fish and other invertebrates that reside in water.

Pollution is also caused when silt and other suspended solids, such as soil, wash off plowed

Words and expressions

prosper ['prɒspə(r)] vi. 繁荣

innocent ['ɪnəsnt] adj. 无辜的

combat ['kɔmbət] vt. 战斗

sewage ['suːɪdʒ] n. 污水
nutrient ['njuːtriənt] n. 营养
phosphate ['fɔsˌfeɪt] n. 磷酸盐

nitrate ['naɪˌtreɪt] n. 硝酸盐
clog [klɒg] vt. 填满（管子）

aquatic [ə'kwætɪk] a. 水生的
respiration [ˌrespə'reɪʃn] n. 呼吸
invertebrate [ɪn'vɜːtəbrɪt]
adj. 无脊椎动物
silt [sɪlt] n. 淤泥

fields, construction and logging sites, urban areas, and crowed riverbanks when it rains. Under natural conditions, lakes, rivers, and other water bodies undergo. Eutrophication is an aging process that slowly fills in the water body with sediment and organic matter. When these enter various bodies of water, fish respiration becomes impaired, plant productivity and water depth become reduced, and aquatic organisms and their environments become suffocated. Pollution in the form of organic material enters waterways in many different forms as sewage, as leaves and grass clippings, or as runoff livestock feedlots and pasture. When natural bacteria and protozoan in the water break down this organic material, they begin to use up the oxygen dissolved in the water. Many types of fish and bottom-dwelling animals cannot survive when levels of dissolved oxygen drop below two to five parts per million. When this occurs, it kills aquatic organisms in large number that leads to disruption in the food chain.

plow [plau] v. 耕地

suffocate ['sʌfə,keɪt] vt. 使窒息

livestock ['laivstɔk] n. 家畜
feedlot ['fi:dlɔt] n. 饲养场
pasture ['pɑ:stʃə] n. 牧场
protozoan [,prəutəu'zəuən] n. 原生动物
bottom-dwelling 底栖生物

Classifying Water Pollution

The major sources of water pollution can be classified as municipal, industrial, and agricultural. Municipal water pollution consists of wastewater from homes and commercial establishments. For many years, the main goal of treating municipal wastewater was simply to reduce its content of suspended solids, oxygen-demanding materials, dissolved inorganic compounds, and harmful bacteria. In recent years, however, more stress has been placed on improving means of disposal of the solid residues from the municipal treatment processes. The basic methods of treating municipal wastewater fall into three stages: primary treatment, including grit removal, screening, grinding, and sedimentation; secondary treatment, which entails oxidation of dissolved organic matter by means of using biologically active sludge, which is then filtered off; and tertiary treatment, in which advanced biological methods of nitrogen removal and chemical and physical methods such as granular filtration and activated carbon absorption are employed. The handling and disposal of solid residues can account for 25 to 50 percent of the capital and operational costs of a treatment plant.

municipal [mju:'nisipl] adj. 都市的

grit [grɪt] n. 砂粒硬渣

filter ['filtə] v. 过滤

granular ['grænjələ] adj. 粒状的

The characteristics of industrial wastewaters can differ considerably both within and among industries. The impact of industrial discharges depends not only on their collective characteristics, such as biochemical oxygen demand and the amount of suspended solids, but also on their content of specific inorganic and organic substances. Three options are available in controlling industrial wastewater. Control can take place at the point of generation in the plant; wastewater can be pretreated for discharge to municipal treatment sources; or wastewater can be treated completely at the plant and either reused or discharged directly into receiving waters.

suspend [sə'spɛnd] vi. 悬浮

Point and Nonpoint Sources

According to the American College Dictionary, pollution is defined as: "to make foul or unclean dirty." Water pollution occurs when a body of water is adversely affected due to the addition of large amounts of material to the water. When it is unfit for its intended use, water is considered polluted.

foul [faʊl] adj. 污秽的

Two types of water pollutants exist: point source and nonpoint source. Point source of pollution occurs when harmful substances are emitted directly into a body of water. The Exxon Valdez oil spill best illustrates point source water pollution. A nonpoint source delivers pollutants indirectly through environmental changes. An example of this type of water pollution is when fertilizer from a field is carried into a stream by rain, in the form of runoff that in turn affects life. The technology exists for point sources of pollution to be monitored and regulated, although political factors may complicate matters. Nonpoint sources are much more difficult to control. Pollution arising from nonpoint sources accounts for majority of the contaminants in streams and lakes.

The Exxon Valdez Oil Spill 瓦尔迪兹漏油事件

Comprehension

Answer the following questions.

1. What is not the reason of water pollution? (　　)
   A. sewage and fertilizers contain nutrients such as nitrates and phosphates
   B. excessive fishes use up the oxygen dissolved in the water
   C. polluted water come to the river from the factory
   D. eutrophication

2. Which kind of water pollution belongs to municipal water pollution?（　　）
   A. agricultural wastewater
   B. plant wastewater
   C. industrial discharges
   D. commercial wastewater
3. Biologically active sludge is used in the（　　）stage of wastewater treatment?
   A. primary　　　B. secondary　　　C. tertiary　　　D. last
4. Which statement is not true according to the text?（　　）
   A. Water is essential for everything on our planet to grow and prosper.
   B. Nitrates and phosphates contribute to excess growth of aquatic plants and algae.
   C. At high BOD levels, organisms that are more tolerant of lower dissolved oxygen may become numerous.
   D. Point sources are much more difficult to control.

Vocabulary building

Active words

organic
adj. 有机的

suspend
vt. 延缓，推迟；停职
vi. 悬浮

subsequent
adj. 后来的

excess
n. 超额
adj. 额外的，过量的

undergo
vt. 经历

content
n. 内容；含量
vt. 使满足

emit
vt. 发射，散发

stream
v. 流动
n.（水、气）流

Useful expressions

parts per million　百万分之一
place stress on　强调；带来压力

account for　解释说明；占

Exercise

Fill in each blank with a given word or expression in their right form.

suspend　　excess　　content　　subsequent　　undergo　　account

1. The planes have to _____ rigorous safety checks.
2. A life without success has no _____.
3. Polyunsaturated oils are essential for health. _____ is harmful, however.
4. The book was banned in the US, as were two _____ books.
5. I'd like to check the balance in my _____ please.
6. Balloons _____ easily in the air.

Extension

Drinking water quality standards

Drinking water quality standards describes the quality parameters set for drinking water. Despite the truism that every human on this planet needs drinking water to survive and that water may contain many harmful constituents, there are no universally recognized and accepted international standards for drinking water.

Many developed countries specify standards to be applied in their own country. In Europe, this includes the European Drinking Water Directive and in the USA the United States Environmental Protection Agency (EPA) establishes standards as required by the Safe Drinking Water Act. For countries without a legislative or administrative framework for such standards, the World Health Organisation publishes guidelines on the standards that should be achieved. China adopted its own drinking water standard GB 3838—2002 (Type II) enacted by Ministry of Environmental Protection in 2002.

The table 5.1 provides a comparison of a selection of parameters for concentrations listed by WHO, the European Union, United States, and Ministry of Environmental Protection of China.

Table 5.1 A Comparison of parameters

Parameter	WHO	European Union	United States	China
Acrylamide		0.10μg/L		
Arsenic	10μg/L	10μg/L	10μg/L	50μg/L
Antimony		5.0μg/L	6.0μg/L	
Barium	700μg/L		2mg/L	
Benzene	10μg/L	1.0μg/L	5μg/L	
Benzo (a) pyrene		0.010μg/L	0.2μg/L	0.0028μg/L
Boron	2.4mg/L	1.0mg/L		
Bromate		10μg/L	10μg/L	
Cadmium	3μg/L	5μg/L	5μg/L	5μg/L
Chromium	50μg/L	50μg/L	0.1mg/L	50μg/L (Cr6)
Copper		2.0mg/L	TT	1mg/L
Cyanide		50μg/L	0.2mg/L	50μg/L
1,2-dichloroethane		3.0μg/L	5μg/L	
Epichlorohydrin		0.10μg/L		
Fluoride	1.5mg/L	1.5mg/L	4mg/L	1mg/L
Lead		10μg/L	15μg/L	10μg/L
Mercury	6μg/L	1μg/L	2μg/L	0.05μg/L
Nickel		20μg/L		
Nitrate	50mg/L	50mg/L	10mg/L (as N)	10mg/L (as N)
Nitrite		0.50mg/L	1mg/L (as N)	
Pesticides (individual)		0.10μg/L		

Continued

Parameter	WHO	European Union	United States	China
Pesticides—Total		0.50μg/L		
Polycyclic aromatic hydrocarbons 1		0.10μg/L		
Selenium	40μg/L	10μg/L	50μg/L	10μg/L
Tetrachloroethene and Trichloroethene	40μg/L	10μg/L		

Exercise

Do you know how to determine the parameters of water? The following table provides some methods for examination. Try to translate it into Chinese.

Parameter	Methods for Examination
Temperature	Thermometer
pH	Electrometric pH meter
Dissolved oxygen	Azide modification
Coliform bacteria	Multiple fermentation technique
Cu	Atomic absorption-direct aspiration
Phenol	Distillation, 4-aminoantipyrene
CN	Pyridine-barbituric acid
DDT	Gas-chromatography

Text B Biological Oxygen Demand Monitoring

Background Information

Microorganisms such as bacteria are responsible for decomposing organic waste. When organic matter such as dead plants, leaves, grass clippings, manure, sewage, or even food waste is present in a water supply, the bacteria will begin the process of breaking down this waste. When this happens, much of the available dissolved oxygen is consumed by aerobic bacteria, robbing other aquatic organisms of the oxygen they need to live. Biological Oxygen Demand (BOD) is a measure of the oxygen used by microorganisms to decompose this waste. If there is a large quantity of organic waste in the water supply, there will also be a lot of bacteria present working to decompose this waste. In this case, the demand for oxygen will be high (due to all the bacteria) so the BOD level will be high. As the waste is consumed or dispersed through

Words and expressions

bacteria [bæk'tɪrɪə] n. [微] 细菌

organic matter 有机物
manure [mə'nʊr] n. 肥料
sewage ['suɪdʒ] n. 污水

aerobic [eə'rəʊbɪk] adj. 需氧的
aquatic [ə'kwætɪk] adj. 水生的

quantity ['kwɒntiti] n. 数量

consume [kən'sjuːm] vt. 消耗

the water, BOD levels will begin to decline.

Nitrates and phosphates in a body of water can contribute to high BOD levels. Nitrate and phosphates are plant nutrients and can cause plant life and algae to grow quickly. When plants grow quickly, they also die quickly. This contributes to the organic waste in the water, which is then decomposed by bacteria. This results in a high BOD level.

When BOD levels are high, dissolved oxygen (DO) levels decrease because the oxygen, which is available in the water, is being consumed by the bacteria. Since less dissolved oxygen is available in the water, fish and other aquatic organisms may not survive.

Test Procedure

The BOD test takes 5 days to complete and is performed using a dissolved oxygen test kit. The BOD level is determined by comparing the DO level of a water sample taken immediately with the DO level of a water sample that has been incubated in a dark location for 5 days. The difference between the two DO levels represents the amount of oxygen required for the decomposition of any organic material in the sample and is a good approximation of the BOD level.

Take 2 samples of water and record the DO level (ppm) of one immediately using the method described in the dissolved oxygen test. Place the second water sample in an incubator in complete darkness at 20℃ for 5 days. If you don't have an incubator, wrap the water sample bottle in aluminum foil or black electrical tape and store in a dark place at room temperature (20℃ or 68℉). After 5 days, take another dissolved oxygen reading (ppm) using the dissolved oxygen test kit. The BOD level is determined by subtracting the Day 5 reading from the Day 1 reading. Record your final BOD results in ppm.

What to Expect

A BOD level of 1-2 ppm is considered very good. There will not be much organic waste present in the water supply. A water supply with a BOD level of 3-5 ppm is considered moderately clean. In water with a BOD level of 6-9 ppm, the water is considered somewhat polluted because there is usually organic matter present and bacteria are

algae [ˈældʒiː] n. 水藻

test kit 试剂盒

incubate [ˈɪŋkjubet] vt. 培养

approximation [əˌprɑksəˈmeʃən] n. 近似值

wrap [ræp] vt. 包

subtract [səbˈtrækt] vt. 减去

decomposing this waste. At BOD levels of 100ppm or greater, the water supply is considered very polluted with organic waste.

Generally, when BOD levels are high, there is a decline in DO levels. This is because the demand for oxygen dissolved in the water. If there is no organic waste present in the water, there won't be as many bacteria present to decompose it and thus the BOD will tend to be lower and the DO level will tend be higher.

At high BOD levels, organisms that are more tolerant of lower dissolved oxygen (i.e. leeches and sludge worms) may appear and become numerous. Organisms that need higher oxygen levels (i.e. caddis fly larvae and mayfly nymphs) will not survive.

tolerant ['tɑlərənt] adj. 耐受的

Adapted from "Bradshaw. A. D. The Treatment and Handling of Waste, 1991".

Comprehension

Choose the best answer according to the text.

1. What does "decompose" mean in paragraph 1?
   A. break out
   B. break through
   C. break down
   D. break away

2. If there is a large quantity of organic waste in the water supply, how will the BOD levels change?
   A. remain original level
   B. begin to decrease
   C. begin to increase
   D. uncertain

3. Which of the follow statements is not true?
   A. If the dissolved oxygen in the river gets less and less, fish and other aquatic organisms will die.
   B. The temperature keeps unchanged during the 5 days in BOD test.
   C. The BOD level is determined by subtracting the Day 5 reading from the Day 1 reading.
   D. Nitrates and phosphates contribute to high dissolved oxygen level.

4. What is the attitude of the author if the BOD levels get higher?
   A. indifference
   B. optimistic
   C. pessimistic
   D. cannot know from the passage

5. What can we infer from the passage?
   A. The change of BOD levels have nothing to do with DO levels.
   B. A COD level of 1-2 ppm is considered very good.
   C. Organisms that need higher oxygen levels will not survive at low BOD.
   D. If BOD levels get higher, the DO levels will be lower.

Vocabulary building

Active words

biological
adj. 生物的；生物学的

oxygen
n. 氧；氧气

demand
n. 要求；需求
v. 需要

decompose
v. 分解，拆解

aerobic
adj. 需氧的

nutrient
n. 营养物

phosphate
n. 磷酸盐

subtract
v. 减

tolerant
adj. 宽容的；有耐药力的

aquatic
adj. 水生的

somewhat
adv. 有点

Useful expressions

be responsible for 是……的缘由；为……负责
break down 分解；发生故障
compare with 与……比较
subtract from 减去

Exercise

Fill in each blank with a given word or expression in their right form.

responsible for   break down   represented as   result from
break out   contribute to

1. This enzyme actively _____ fat compounds and lipids.
2. Death usually _____ hepatic or renal failure
3. His carelessness _____ the accident.
4. Mike is _____ designing the entire project.
5. He was 29 when war _____ .
6. The king is _____ a villain（反面人物）in the play.

Extension

Monitoring Method

NO_2 are measured continuously by the principle of chemilum inescences. In this method, the air sample is split into two pathways; one to measure NO, and the other to measure total NO_2. In the first pathway, the sample is mixed with O_3 and light is produced. The amount of light is proportional to the NO concentration. In the second pathway, a catalytic converter is used to change all the NO_2 to NO, and the NO is measured again, The sum of the readings from the first and second pathways is equal to the total NO_2; the difference of the readings is the NO_2 concentration.

CO is monitored continuously by either non-depressive infrared photometry or gas filter correlation. The non-depressive infrared process is based upon the absorption of infrared light by CO. Gas filter correlation is operated on the same principle as non-depressive infrared photometry, but is more specific to CO by eliminating water vapors, CO_2 and other interference.

Unlike other pollutants, ozone (O_3) is not emitted directly by human activities. O_3 in the lower atmosphere is produced by a complicated set of chemical reactions involving oxides of nitrogen (NO_2) and volatile organic compounds (VOCs) in the presence of sunlight. O_3 is also transported to the ground from the "ozone rich" upper atmosphere by natural weather recesses. O_3 and its precursors, such as NO_2 and VOCs, may also be carried from upwind sources such as urban centers and industrial complexes. A major source of VOCs in rural areas is natural emissions from trees and vegetation.

O_3 concentrations are generally lower at urban locations than at rural locations. This is due to the destruction of O_3 by nitric oxide (NO) that is emitted by vehicle. O_3 levels are usually higher during the spring and summer months because of more transportation from the upper atmosphere and more sunlight which allow O_3 forming chemical reactions to occur more rapidly. O_3 is a colorless, odorless gas. However, O_3 does have a characteristic of sharp odors when at very high concentrations, such as that associated with lightning storms.

Text C Recycling and Reuse

Words and expressions

Recycling involves the collection of the used and discarded materials, processing these materials and making them into new products. It reduces the amount of waste that is thrown into the community dustbins thereby making the environment cleaner and the air fresher to breathe.

discard [dɪsˈkɑrd] vt. 丢弃

community [kəˈmju:nɪti] n. 社区
dustbin [ˈdʌstbɪn] n. 垃圾箱

Surveys carried out by government and non-government agencies in the country have all recognized the importance of recycling wastes. However, the methodology for safe recycling of waste has not been standardized. Studies have revealed that 7%-15% of the waste is recycled. If recycling is done in a proper manner, it will solve the problems of waste or garbage. At the community level, a large number of NGOs (Non-Governmental Organizations) and private sector enterprises have taken an initiative in segregation and recycling of waste. It is being used for composting, making pellets to be used in gasifies, etc. Plastics are sold to the factories that

survey [ˈsɜve] n. 调查

recognize [ˈrekəgnaɪz] v. 确认
methodology [ˌmɛθəˈdɑlədʒi] n. 方法
reveal [rɪˈvi:l] vt. 显示

initiative [ɪˈnɪʃətɪv] adj. 主动的
segregation [ˌsɛgrɪˈgeʃən] n. 隔离
compost [ˈkɑmpɔst] n. 堆肥

reuse them. Some items that can be recycled are shown in the table 5.2.

Table 5.2 Some Items that Can Be Recycled or Reused

Paper	old copies old books paper bags newspaper old greeting cards
Plastic	containers bottles bags sheets
Glass and Ceramics	bottles plates cup bowls
Miscellany	old cans utensils clothes furniture

ceramic [sə'ræmɪk] n. 陶瓷
bowl [boʊl] n. 碗

miscellany ['mɪsəleni] n. 杂类
utensil [jʊ'tɛnsl] n. 器皿

The steps involved in the process prior to recycling include:

1. Collection of waste from doorsteps, commercial place.

2. Collection of waste from community dumps.

3. Collection/picking up of waste from final disposal sites.

Most of the garbage generated in the household can be recycled and reused. Organic kitchen waste such as leftover foodstuff, vegetable peel, spoilt or dried fruit and vegetable can be recycled by putting them in the compost pits, which have been dug in the garden. Old newspaper, magazines and bottles can be sold to the man who buys these items from homes.

In your own homes you can contribute to waste reduction and the recycling and reuse of certain items. To cover your books, you can use old calendars. Old greeting cards can also be reused. Paper can also be made at home through a very simple process and you can paint on them.

Waste recycling has some significant advantages:

1. Lead to less utilization of raw materials.

2. Reduce environmental impacts arising from waste treatment and disposal.

3. Make the surroundings cleaner and healthier.

doorstep ['dɔːstep] n. 门阶

dumps [dʌmps] n. 垃圾场
disposal site 废物处理场

compost pit 堆肥坑

rag-picker 拾荒者
junk dealer 废品商

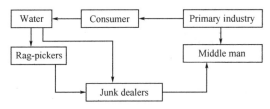

Fig. 5.1   Recycling of wastes
(Source: management of municipal solid wastes.)

4. Save on landfill space.
5. Save money.
6. Reduce the amount of energy required to manufacture new products.

In fact recycling can prevent the creation of waste at the source.

landfill ［lændfɪl］ n. 垃圾填埋地

Adaptd from "Joseph L. Pavoni and John E. Handbook of solid Waste Disposal, Van Reinhold Company, New York, 1975".

Comprehension

Choose the best answer according to the text.

1. Which statement does not belong to "recycle and reuse"?
   A. Cover your books with old calendars.
   B. Collection of waste from community dumps.
   C. Treat things that have already been used so that they can be used again.
   D. Organic kitchen waste is fermented and then used as fertilizer in the garden.
2. Which substance can be recycled or reused?
   A. miscellany   B. paper and plastic
   C. glass and ceramic   D. A，B and C
3. How many steps involved in the process prior to recycling?
   A. five   B. eight   C. three   D. ten
4. What is the attitude of the author about the "recycle and reuse"?
   A. disagree   B. agree   C. positive   D. negative

Vocabulary building

Active words

material
n. 材料
adj. 物质的

product
n. 产品，产物；乘积

standardized
adj. 标准的

reveal
v. 揭露，揭发

private
adj. 私人的；个人的；私下的；私有的；缄默的
n. 士兵；列兵

recycle
v. 重复利用

Useful expressions

prior to 在……之前   save on 节省
arise from 产生

Exercise

Fill in each blank with a given word or expression in their right form.

recycle   level   contribute   reduce   impact   generate

1. The standard of your work has fallen from the _____ we expect from you.
2. They are selling the goods under the counter at _____ prices.
3. They _____ empty tins so as to use the metal.
4. He has made an important _____ to the company's success.
5. This will have almost no _____ on battery life.
6. Big uranium atoms are split into smaller atoms. That _____ heat plus neutrons.

Extension

Preferred handling options for common products is shown in table 5.3.

Table 5.3 Preferred handling options for common products

Product type	Personal safety	Options
HOUSEHOLD		
Abrasive Cleaner	May contain ammonia (see ammonia)	U
Ammonia	DO NOT mix with bleach (poisonous gas)	U
Bleach	DO NOT mix with ammonia or acids (poisonous gas)	U
Disinfectant	May contain bleach (see bleach)	U
Drain Opener	May contain lye (corrosive)	U A
Flea Collar	Avoid skin contact	U A
Furniture Polish	Keep away from heat & flame	U A
Household Batteries	Beware of leakage corrosive	R A
Mothballs	Keep away from children & pets (resembles candy)	U A
Mouse & Rat Poison	Keep out of reach of children & pets	U A
Furniture Oils	Don't bury oily rags in a rag bucket, clothes hamper, etc. (may spontaneously combust if not allowed to air out)	U A
Oven Cleaner	May contain lye (corrosive)	U A
Roach & Ant Killer	Keep out of reach of children & Pets	U A
Rug & Upholstery Cleaner	Avoid skin contact	U A
WORKSHOP		
Enamel or Oil-Based Paint	Keep away from heat & flame	U R A
Latex or Water-Based Paint	May contain mercury (toxic)	U R A
Paint Thinner	Avoid skin contact	U A

Continued

Product type	Personal safety	Options
Paint & Varnish Remover	Keep away from heat & flame	U A
Photographic Chemicals	Use exhaust hood and avoid skin contact	U A
Stains & Varnishes	Keep away from heat & flame	U A
AUTO		
Antifreeze	Keep away from children & pets (highly toxic and sweet taste)	U R A
Auto Battery	Beware of leakage; corrosive	U R A
Used Oil	Avoid prolonged exposure	R A
Transmission Fluid	Avoid prolonged exposure	R A
Windshield Wiper Fluid	May contain methanol (toxic)	U A
LAWN		
Fertilizer	Keep separated from fuel oil, gasoline, etc.	U A
Insecticides	Avoid inhalation, skin contact	U A
Herbicides	Avoid inhalation, skin contact	U A

U Use Up   R Recycle   A Save for HHW Collection

All hazardous products should be: kept out of reach of children & pets; used in well-ventilated areas only; and kept in original containers & labelled. Always thoroughly wash your hands after handling hazardous products.

Adapted from http://environmentalchemistry.com/

Reading material

Air Pollution

Air is the ocean we breathe. Air supplies us with oxygen which is essential for our bodies to live. Air is 99.9% nitrogen, oxygen, water vapor and inert gases. Human activities can release substances into the air, some of which can cause problem for humans, plants, and animals.

Air pollution is normally defined as air that contains one or more chemicals in high enough concentrations to harm humans, other animals, vegetation or materials. There are two major types of air pollutants. A primary air pollutant is a chemical added directly to the air that occurs in a harmful concentration. It can be a natural air component, such as carbon dioxide that rises above its normal concentration or something not usually found in the air, such as a lead compound emitted by cars burning leaded gasoline. A secondary air pollutant is a harmful chemical formed in the atmosphere through a chemical reaction among air components. Serious air pollution usually results over a city or other area that is emitting high levels of pollutants during a period of air stagnation. The geographic location of some heavily populated cities, such as Los Angeles and Mexico City, makes them particularly susceptible to frequent air stagnation and pollution buildup.

There are several main types of pollution and well-known effects of pollution, which are commonly discussed. These include smog, acid rain, the greenhouse effect, and "holes" in the ozone layer. Such of these problems has serious implications for our health as well as for the whole environment.

One type of air pollution is the release of particles into the air from burning fuel for energy. Diesel smoke is a good example of this particulate matter. The particles are very small pieces of matter measuring about 2.5 microns or about 0.0001 inches. This type of pollution is sometimes referred to as black carbon pollution The exhaust from burning fuels in automobiles, homes, and industries is a major source of pollution in the air. Some authorities believe that even the burning of wood and charcoal in fireplaces and barbeques can release significant quantities of soot into the air.

Another type of pollution is the release of noxious gases, such as sulfur dioxide, carbon monoxide, nitrogen oxides, and chemical vapors. These can take part in further chemical reactions once they are in the atmosphere, forming smog and acid rain.

Pollution also needs to be considered inside our homes, offices, and schools. Some of these pollutants can be created by indoor activities such as smoking and cooking. In the United States, People spend about 80%-90% of our time inside the building, so their exposure to harmful indoor pollutants can be serious. It is therefore important to consider both indoor and outdoor air pollution.

Outdoor Air Pollution

Smog is a type of large-scale outdoor pollution. It is caused by chemical reactions between pollutants derived from different sources, primarily automobile exhaust and industrial emissions. Cities are often centers of these types of activities, and many suffer from the effects of smog, especially during the warm months of the year.

For each city, the causes of pollution may be different. Depending on the geographical location, temperature, wind and weather factor, pollution is dispersed differently. However, sometimes this does not happen and the pollution can build up to dangerous levels. A temperature inversion occurs when air close to the earth is cooler than the air above it. Under these conditions, the pollution cannot rise and be dispersed. Cities surrounded by mountains also experience trapping on pollution. Inversion can happen in any season. Winter inversions are likely to cause particulate and carbon monoxide pollution. Summer inversions are more likely to create smog.

Another consequence of outdoor air pollution is acid rain. When a pollution, such as sulfuric acid, combines with droplets of water in the air, the water (or snow) can become acidified. The effects of acid rain on the environment can be very serious. It damages plants by destroying their leaves, it poisons the soil, and it changes the chemistry of lakes and streams. Damage due to acid rain kills trees and harms animals. US Geological Survey (USGS), the Environmental Protection Agency (EPA), and Environment Canada are among the organizations that are actively studying the acid rain problem.

The greenhouse effect, also referred to as global warming, is generally believed to come from the build up of carbon dioxide gas in the atmosphere. Carbon dioxide is produced when fuels are burned. Plants convert carbon dioxide back to oxygen, but the release of carbon dioxide from human activities is higher than the amount that plants can process. The situation is made worse since many of the earth's forests are being removed, and plant life is being damaged by acid rain. Thus, the amount of carbon dioxide in the air is continuing to increase. This build up acts like a blanket and traps heat close to the surface of our earth. Changes of even a few degrees will affect us all through the changes in the climate and even the possibility that the polar ice caps may melt. (One of the consequences of polar ice cap melting would be sea level rise, resulting in widespread coastal flooding.)

Ozone depletion is another result of pollution. Chemicals released by our activities affect the stratosphere, one of the atmospheric layers surrounding earth. The ozone layer in the stratosphere protects the earth from harmful ultraviolet radiation from the sun. Release of chlorofluorocarbons (CFC's) from aerosol cans, cooling systems and refrigerator equipment removes some of the ozone, causing "holes"; to open up in this layer and allowing the radiation to reach the earth. Ultraviolet radiation is known to cause skin cancer and has damaging effects on plants and wildlife.

Indoor Air Pollution

Many people spend large portion of time indoors as much as 80%-90% of their lives. We work, study, eat, drink and sleep in enclosed environments where air circulation may be restricted. For these reasons, some experts feel that more people suffer from the effects of indoor air pollution than outdoor pollution.

There are many sources of indoor air pollution. Tobacco smoke, cooking and heating appliances, and vapors from building materials, paints, furniture, etc. cause pollution inside buildings. Radon is a natural radioactive gas released from the earth, and it can be found concentrated in basements in some parts of the United State. Pollution exposure at home and work is often greater than outdoors. The California Air Resources Board estimates that indoor air pollutant levels are 25%-62% greater than outside levels and can pose serious health problem.

Comprehension

Answer the following questions according to the passage.

1. What's the composition of air?
2. What are two major types of outdoor air pollutants?
3. What pollutions do winter invertions cause?
4. According to The California Air Resources Board, are indoor air pollutant levels greater than outside levels?

Supplementary knowledge

Structure of a Journal-Style Scientific Paper

Most journal-style scientific papers are subdivided into the following sections: Title, Authors and Affiliation, Abstract, Introduction, Methods, Results, Dicussion, Acknowledgements, and Literature Cited, which parallel the experimental process. The sections appear in a journal style paper is shown in table 5.4.

Table 5.4 Sections in a Journal style paper

Experimental process	Section of Paper
What did I do in a nutshell?	Abstract
What is the problem?	Introduction
How did I solve the problem?	Materials and Methods
What did I find out?	Results
What does it mean?	Discussion
Who helped me out?	Acknowledgments (optional)
Whose work did I refer to?	Literature Cited
Extra Information	Appendices (optional)

1. ABSTRACT

An abstract summarizes, in one paragraph (usually), the major aspects of the entire paper in the following prescribed sequence:
- the question (s) you investigated (or purpose);
- the experimental design and methods used;
- the major findings including key quantitative results, or trends;
- a brief summary of your interpretations and conclusions.

2. INTRODUCTION

The function of the Introduction is to:
- Establish the context of the work being reported. This is accomplished by discussing the relevant primary research literature (with citations) and summarizing our current understanding of the problem you are investigating;
- State the purpose of the work in the form of the hypothesis, question, or problem you investigated; and,
- Briefly explain your rationale and approach and, whenever possible, the possible outcomes your study can reveal.

Use the active voice as much as possible. Some use of first person is okay, but do not overdo it.

3. MATERIALS AND METHODS

In this section you explain clearly how you carried out your study in the following general structure and organization (details follow below):
- the *the organism (s) studied* (plant, animal, human, etc.) and, when relevant, their pre-experiment handling and care, and when and where the study was carried out (only if location and time are important factors);
- if you did a field study, provide a description of the study site, including the significant physical and biological features, and the precise location (latitude and longitude, map, etc.);
- the experimental OR sampling design [i.e., how the experiment or study was structured. For example, controls, treatments, what variable (s) were measured, how many samples were collected, replication, the final form of the data, etc.];
- the protocol for collecting data, i.e., how the experimental procedures were carried out; and,

• how the data were analyzed (qualitative analyses and/or statistical procedures used to determine significance, data transformations used, what probability was used to decide significance, etc.).

4. RESULTS

The function of the Results section is to objectively present your key results, without interpretation, in an orderly and logical sequence, using both text and illustrative materials (tables and figures). The results section always begins with text, reporting the key results and referring to your figures and tables as you proceed. Summaries of the statistical analyses may appear either in the text (usually parenthetically) or in the relevant tables or figures (in the legend or as footnotes to the table or figure). The results section should be organized around tables and/or figures that should be sequenced to present your key findings in a logical order. The text of the results section should be crafted to follow this sequence and highlight the evidence needed to answer the questions/hypotheses you investigated. Important negative results should be reported, too.

5. DISCUSSION

Fundamental questions to answer here include:

• Do your results provide answers to your testable hypotheses? If so, how do you interpret your findings?

• Do your findings agree with what others have shown? If not, do they suggest an alternative explanation or perhaps a unforeseen design flaw in your experiment (or theirs)?

• Given your conclusions, what is our new understanding of the problem you investigated and outlined in the Introduction?

• If warranted, what would be the next step in your study, e.g., what experiments would you do next?

6. ACKNOWLEDGMENTS

If, in your experiment, you received any significant help in thinking up, designing, or carrying out the work, or received materials from someone who did you a favor by supplying them, you must acknowledge their assistance and the service or material provided. Authors always acknowledge outside reviewers of their drafts (in PI courses, this would be done only if an instructor or other individual critiques the draft prior to evaluation) and any sources of funding that supported the research. Although usual style requirements (e.g., 1st person, objectivity) are relaxed somewhat here, acknowledgments are always brief and never flowery.

7. LITERATURE CITED

The literature cited section gives an alphabetical listing (by first author's last name) of the references that you actually cited in the body of your paper.

8. APPENDICES

An appendix contains information that is non-essential to understanding of the paper, but may present information that further clarifies a point without burdening the body of the presentation. An appendix is an optional part of the paper, and is rarely found in published papers.

Adapted from http://abacus.bates.edu

Marketing & Selling

Objectives:

After finishing this module, you are able to:
- Negotiate the chemical product with business partners properly
- Write some certain business letters and e-mails
- Get to know how to introduce a company or some certain kind of product in English

Warming-up

Which Chemical Firm will be Your Target? Why?

Global Top 50

The world's largest CHEMICAL FIRMS are growing and enjoying stronger profits.

GLOBAL TOP 50

BASF retained the lead, but Sinopec overtook Dow Chemical to claim the number two spot

RANK 2013	RANK 2012	COMPANY	CHEMICAL SALES ($ MILLIONS) 2013	CHANGE FROM 2012	CHEMICAL SALES AS % OF TOTAL SALES	HEAD-QUARTERS COUNTRY	CHEMICAL OPERATING PROFITS[a] ($ MILLIONS)	CHANGE FROM 2012	CHEMICAL PROFITS AS % OF TOTAL OPERATING PROFITS	OPERATING PROFIT MARGIN[b]	IDENTIFIABLE CHEMICAL ASSETS ($ MILLIONS)	CHEMICAL ASSETS AS % OF TOTAL ASSETS	OPERATING RETURN ON CHEMICAL ASSETS[c]
1	1	BASF	$78,615	-4.6%	80.0%	Germany	$6,317	-6.2%	65.4%	8.0%	$69,676	81.5%	9.1%
2	3	Sinopec	60,829	5.0	13.0	China	103	71.9	0.6	0.2	25,427	12.3	0.4
3	2	Dow Chemical	57,080	0.5	100.0	U.S.	4,715	6.6	100.0	8.3	69,501	100.0	6.8
4	5	SABIC	43,589	3.1	86.5	Saudi Arabia	12,795	1.7	86.7	29.4	84,207	93.1	15.2
5	4	Shell[d]	42,279	-7.6	9.4	Netherlands	na	na	na	na	na	na	na
6	6	ExxonMobil	39,048	0.8	9.3	U.S.	5,180	6.0	9.1	13.3	27,475	7.9	18.9
7	7	Formosa Plastics[e]	37,671	5.9	60.2	Taiwan	2,352	67.2	62.8	6.2	43,060	66.6	5.5
8	8	LyondellBasell Industries	33,405	1.7	75.8	Netherlands	5,087	17.5	99.7	15.2	na	na	na
9	9	DuPont[d]	31,044	2.7	86.9	U.S.	5,234	11.6	97.5	16.9	18,113	66.2	28.9
10	12	Ineos	26,861	-10.8	100.0	Switzerland	2,137	-6.3	100.0	8.0	na	na	na
11	10	Mitsubishi Chemical	26,685	14.8	74.4	Japan	507	121.1	44.8	1.9	23,411	65.7	2.2
12	11	Bayer	26,636	0.9	49.9	Germany	4,409	1.0	39.5	16.6	25,571	37.5	17.2
13	13	LG Chem	21,142	-0.5	100.0	South Korea	1,592	-8.8	100.0	7.5	15,938	100.0	10.0
14	14	AkzoNobel	19,376	-5.2	100.0	Netherlands	1,193	-3.5	100.0	6.2	21,332	100.0	5.6
15	16	Air Liquide	19,153	-0.8	94.7	France	3,569	1.1	96.9	18.6	29,595	95.2	12.1
16	17	Braskem	18,994	15.4	100.0	Brazil	1,370	140.1	100.0	7.2	22,414	100.0	6.1
17	19	Mitsui Chemicals	18,916	11.5	100.0	Japan	306	597.1	100.0	1.6	13,634	100.0	2.2
18	23	Linde	18,554	11.0	83.9	Germany	5,108	13.0	97.0	27.5	na	na	na
19	15	Sumitomo Chemical	18,116	16.3	78.8	Japan	688	136.9	66.6	3.8	18,163	63.6	3.8
20	18	Reliance Industries	17,778	10.4	23.3	India	1,436	17.4	35.2	8.1	9,844	13.4	14.6
21	21	Evonik Industries	17,097	-3.7	100.0	Germany	1,653	-22.5	100.0	9.7	21,113	100.0	7.8
22	20	Toray Industries	16,665	17.9	88.5	Japan	1,152	22.5	106.8	6.9	18,734	86.3	6.1
23	26	Lotte Chemical	15,017	3.4	100.0	South Korea	445	31.1	100.0	3.0	9,763	100.0	4.6
24	24	Yara	14,472	0.6	100.0	Norway	1,963	-23.1	100.0	13.6	15,140	100.0	13.0
25	25	PPG Industries	14,044	-0.9	93.0	U.S.	2,134	-3.0	97.4	15.2	11,900	75.0	17.9
26	22	Solvay	13,768	-19.2	100.0	Belgium	1,179	-24.0	100.0	8.6	24,479	100.0	4.8
27	27	Chevron Phillips	13,147	-1.2	100.0	U.S.	na	na	na	na	10,533	100.0	na
28	30	DSM	12,773	5.3	100.0	Netherlands	580	-11.9	100.0	4.5	15,959	100.0	3.6
29	28	Shin-Etsu Chemical[d]	11,945	13.7	100.0	Japan	1,781	10.7	100.0	14.9	22,530	100.0	7.9
30	32	Praxair	11,925	6.2	100.0	U.S.	3,734	7.9	100.0	31.3	20,255	100.0	18.4
31	34	SK Innovation	11,640	1.4	19.1	South Korea	770	12.2	61.0	6.6	5,502	17.1	14.0
32	29	Asahi Kasei	11,199	15.3	57.6	Japan	632	107.1	38.9	5.6	9,774	50.3	6.5
33	33	Huntsman Corp.	11,079	-1.0	100.0	U.S.	671	-27.9	100.0	6.1	9,188	100.0	7.3
34	31	Lanxess	11,023	-8.7	100.0	Germany	406	-62.4	100.0	3.7	9,045	100.0	4.5
35	38	Borealis	10,815	7.9	100.0	Austria	259	23.4	100.0	2.4	10,230	100.0	2.5
36	37	Syngenta	10,793	5.7	73.5	Switzerland	na	na	na	na	na	na	na
37	36	Sasol	10,225	11.9	54.4	South Africa	199	-70.5	4.7	1.9	9,140	35.7	2.2
38	35	Mosaic	9,974	-10.2	100.0	U.S.	2,333	-12.8	100.0	23.4	18,086	100.0	12.9
39	—	PTT Global Chemical	9,959	6.2	58.5	Thailand	1,595	8.5	83.9	16.0	10,221	72.6	15.6
40	39	Air Products & Chemicals	9,729	5.8	95.6	U.S.	1,518	-4.0	95.9	15.6	16,162	90.5	9.4
41	43	Eastman Chemical	9,350	15.4	100.0	U.S.	1,938	110.7	100.0	20.7	11,845	100.0	16.4
42	42	Arkema	8,098	-4.6	100.0	France	781	-13.3	100.0	9.6	7,272	100.0	10.7
43	41	Tosoh[d]	7,913	15.5	100.0	Japan	426	69.9	99.6	5.4	7,395	100.0	5.8
44	46	Styrolution	7,703	-3.3	100.0	Germany	587	31.9	100.0	7.6	na	na	na
45	40	DIC	7,606	5.5	100.0	Japan	439	11.4	100.0	5.8	6,974	100.0	6.3
46	47	Total[f]	7,570	0.0	3.3	France	584	14.6	4.8	7.7	na	na	na
47	—	Indorama	7,464	8.7	100.0	Thailand	146	-16.1	100.0	2.0	6,159	100.0	2.4
48	45	Eni	7,397	-7.3	4.9	Italy	-963	nm	def	def	4,208	2.9	def
49	44	PotashCorp	7,305	-7.8	100.0	Canada	2,365	-21.5	100.0	32.4	17,958	100.0	13.2
50	48	Alpek	7,059	-6.3	100.0	Mexico	428	-23.9	100.0	6.1	4,556	100.0	9.4

NOTE: Some figures converted at 2013 average exchange rates of $1.00 U.S. = 2.157 Brazilian reals, 6.15 Chinese renminbi, 0.753 euros, 58.51 Indian rupees, 97.6 Japanese yen, 1,094.67 Korean won, 5.8772 Norwegian krone, 3.75 Saudi riyals, 0.9269 Swiss francs, 29.68 new Taiwan dollars, and 30.696 Thai baht. a Operating profit is sales less administrative expenses and cost of sales. b Operating profit as a percentage of sales. c Chemical operating profit as a percentage of identifiable assets. d Sales include a significant amount of nonchemical products. e C&EN estimates. f Chemical figures include only specialty chemicals. **def** = deficit. **na** = not available. **nm** = not meaningful.

Text A BASF and Other Top Chemical Firms

Words and expressions

BASF 巴斯夫股份公司
C&EN《化学工程新闻》杂志
tournament ['tuənəmənt] n. 锦标赛
contender [kən'tendə(r)] n.（冠军）争夺者
hands-down adj. 无疑的
consecutive [kən'sekjətɪv] adj. 连续的
enormous [ɪ'nɔːməs] adj. 巨大的；庞大的
Sinopec 中国石油化工集团公司（全称 Sinopec Group）
sales n. 销售（额）
amount to 共计达到 revenue ['revənjuː] n. 收入
region ['riːdʒən] n. 地区
put a damper on 扫兴

Germany won soccer's World Cup championship final earlier this month. The country also won—or at least Germany's BASF did—C&EN's Global Top 50 ranking of the world's largest chemical producers.

But unlike in the soccer tournament, where Germany was a main contender but not the hands-down favorite, there was little doubt that BASF would come out on top of the C&EN survey. After all, the firm has been there for 9 consecutive years.

BASF is truly an enormous chemical company. Its $78.6 billion in chemical sales for 2013, the year on which the survey is based, is $17.8 billion more than the sales recorded by the second-largest firm, China's Sinopec. The differential is bigger than the sales of the number 20 company in the ranking, India's Reliance Industries. BASF's sales amount to 8.0% of the combined revenue of all of the companies on the list.

Furthermore, BASF is big in every region of the world. Its North American business alone would be number 14 on the global list. Any economic factor that would put a damper on BASF's sales would take most every other large chemical firm down with it.

So BASF will likely be the world's largest chemical company for years to come.

Few acquisitions among large chemical makers would be big enough to dislodge the German firm. And given that Verbund, a German word meaning something like "integration," is a core BASF value, a breakup of BASF isn't likely.

acquisitions [ˌækwɪ'zɪʃn] n. 获得
dislodge [dɪs'lɒdʒ] vt. 驱逐
integration [ˌɪntɪ'greɪʃn] n. 整合

If the Global Top 50 can be considered a competition, it is a contest among the 49 firms that aren't BASF. This group has actually experienced some jostling for position.

jostling ['dʒɒslɪŋ] n. 争抢

This is Sinopec's first year in the number two slot, having edged out Dow Chemical. Sinopec was a close third last year, but a 5.0% increase in sales and a strengthening Chinese renminbi combined to lift the company over Dow, which experienced a paltry 0.5% increase in sales.

slot [slɒt] n. 位置
edge out 排挤
Dow Chemical 美国的陶氏化学公司
renminbi 人民币
paltry ['pɔːltri] adj. 微小的

In his annual letter to shareholders, Sinopec Chairman Chengyu Fu noted that the firm's chemical business "successfully mitigated the impact of difficult market conditions."

As for the future, Fu echoed the kind of optimism that will sound familiar to China watchers. "China's economy will become all the more vibrant as economic reforms allow markets to play a more decisive role in resource allocation," he wrote, "The continuous pursuit of industrialization and urbanization will support steady growth in demand for oil and petrochemical products."

Also at the top of the ranking, Saudi Basic Industries Corp. overtook Shell Chemicals to claim the number four slot. SABIC's 3.1% increase in sales was only a modest improvement, but Shell's sales declined 7.6%.

The Swiss firm Ineos broke into the top 10, but only because of a technicality. In previous years, C&EN counted only the results of Ineos Group Holdings, which comprises mainly its petrochemical and polyethylene businesses. This year, the company provided results that aggregated other operations, such as its polyvinyl chloride business. If C&EN had counted those operations last year, Ineos would have been ranked 10 instead of 12.

mitigate ['mɪtɪgeɪt] vt. 使缓和

vibrant ['vaɪbrənt] adj. 充满生气的
resource allocation 资源分配

Saudi Basic Industries Corp. 沙特基础工业公司
overtake [ˌəʊvə'teik] v. 追上，赶上
Shell Chemicals 皇家荷兰壳牌公司集团
Ineos 英力士集团 comprises [kəm'praɪz] vt. 包含
petrochemical [ˌpetrəʊ'kemɪkl] n. 石油化学产品
polyethylene [ˌpɒli'eθəliːn] n. 聚乙烯
aggregate ['ægrɪgət] vt. 使聚集
polyvinyl chloride n. 聚氯乙烯

Adapted from the cover story of CHEMICAL & ENGINEERING NEWS Volume 92 Issue 30

Comprehension

Choose the best answer according to the text.

1. ( ) won the first place in C&EN's Global Top 50 ranking of the world's largest chemical producers in 2014.
    A. Shell
    B. Ineos
    C. BASF
    D. Sinopec

2. The chemical sales of Sinopec for 2013 was ( ) billion.
    A. $78.6
    B. $17.8
    C. $96.4
    D. $60.8

3. What is the main reason why Sinopec overtook Dow to take the second slot in this year? ( )
    A. A 5.0% increase in sales

B. A strengthening Chinese renminbi
C. China's economic reforms
D. The continuous pursuit of industrialization and urbanization

4. Which statement is not true according to the text? （　　）

A. BASF won the C&EN's Global Top 50 ranking of the world's largest chemical producers with no doubt.
B. Ineos would have been the top 10 last year if counted as the way in this year.
C. BASF in North America wins the 14 slot according to C&EN's Global Top 50 ranking.
D. The top five chemical companies are：BASF，Shell Chemicals，SABIC，Sinopec，Dow Chemical.

Vocabulary building

Active words

consecutive
adj. 连续的，不间断的

enormous
adj. 巨大的；庞大的

sales
n. 销售（额）

revenue
n. 收入；营业额

region
n. 地区，地域

integration
n. 结合；整合

slot
n. 位置

overtake
v. 追上，赶上

comprise
vt. 包含，包括

aggregate
vt. 使聚集，使积聚；总计达
n. 聚集体；集料（可成混凝土或修路等用的）
adj. 总数的，总计的；聚合的

Useful expressions

amount to　共计达到，折合为

edge out　排挤；小胜；险胜

Exercise

Fill in each blank with a given word or expression in their right form.

overtake　　edge out　　sales　　amount to　　revenue　　comprise

1. In 1987，McDonald's captured 19 percent of all fast-food _____.
2. Rent is one form of _____.
3. We are expected to _____ the developed countries in the next century.
4. These _____ asphalt，tar and waxes.
5. There was such a big crowd at the gate that we had to _____.
6. The bill _____ $500.

Extension

The Model of Introducing a Company

××公司成立于××年，地处××，享有便捷的交通和优美的环境。我公司占地面积××平方米，员工人数为××。	_____（公司英文名）was established in _____（公司成立时间）and is located in _____（公司所在城市），enjoying convenient transportation access and a beautiful environment. Our company covers an area of _____（占地面积）square meters and has _____（员工人数）employees.
我们主要从事××，年产量达到××。我公司在××行业有丰富的经验，主要产品包括××。除此之外，我们还致力于开发新产品以满足不同客户的不同要求。我们有××系统来严格控制产品的质量。	We are specialized in _____（主营业务）and have an annual production capacity of _____（年产量）. Our company has rich experience in the _____（所从事的行业）industry. Our main products include _____（主要产品）. Besides, we are making great efforts to develop new products to meet different requirements from different customers. Our company has _____（质量控制系统）to strictly control product quality.
我们坚持互惠互利原则，在我们的客户中享有良好声誉，因为我们提供完美的服务，高质量的产品以及有竞争力的价格。热忱欢迎国内外客户与我们合作来获得共同的成功。	Adhering to the business principle of mutual benefit, we have built up a good reputation among our customers due to our perfect services, quality products and competitive prices. We warmly welcome customers at home and abroad to cooperate with us for the common success.

Practice

Try to introduce BASF to one of your friends, and write down what you want to say. If there is something unknown about BASF, try to find it out via website.

_____
_____
_____
_____
_____
_____
_____
_____
_____

Text B Inquires

Dear Sir/Madam,

Thanks for your mail and we would appreciate that if you always send the following details while making correspondences with us.

1) Price
2) COA and MSDS
3) Payment terms
4) Load port
5) Your bank details
6) The name of companies in India to whom you have supplied in recent times.

The above information will always help us to minimize the un-necessary correspondences.

In this connection please note that we buy following chemicals from China.

- Ammonium Chloride (99.5%) Industry/Food Grade
- Ammonium Bicarbonate
- Acetic Acid Glacial
- Calcium Lignosulphonate
- Caustic Soda Flakes
- Cyclohexanone
- Citric Acid Monohydrate/Anhydrate
- Glycerine
- Gum rosin WW Grade
- Hydrogen Peroxide 50%
- Melamine
- Phosphoric Acid Tech/Food Grade
- Soda Ash Light
- Soda Ash Dense
- Sodium Gluconate
- Sodium Hyposulfite 85%/88%
- Sodium Sulphide Yellow Flakes (30/60ppm)

We request you please give us a quote on CIF Kolkata basis in 2-3 FCLs on each items. Please note that our payment term is L/C 90 days only and we have been maintaining this payment term since last 20 years with China.

Kindly acknowledge and look forward to your reply.

Words and expressions

appreciate [ə'priːʃieɪt] vt. 感激；欣赏；领会
correspondence [ˌkɒrə'spɒndəns] n. 通信，信件
COA Certificate of Analysis 分析单
MSDS Material Safety Data Sheet 化学品安全技术说明书
payment term 支付条件、付款条件
load port 装货港
supply [sə'plaɪ] vt. 供给；向……提供（物资等）
minimize ['mɪnɪmaɪz] vt. 把……减至最低数量（或程度）
ammonium [ə'məʊniəm] n. 铵
chloride ['klɔːraɪd] n. 氯化物
industry/food grade 工业/食品级
bicarbonate [ˌbaɪ'kɑː bənət] n. 重碳酸盐
acetic [ə'siːtɪk] adj. 乙酸的
glacial ['gleɪʃl] adj. 冰的
calcium lignosulphonate 木质素磺酸钙
caustic soda 苛性钠
flake [fleɪk] n. 小薄片
cyclohexanone [ˌsaɪkləʊ'heksənəʊn] n. 环己酮
citric acid 柠檬酸
anhydrate [æn'haɪdreɪt] v. 脱水
glycerine ['glɪsəriːn] n. 甘油
gum rosin [gʌm'rɒzɪn] n. 松香
hydrogen peroxide 过氧化氢
melamine ['meləmiːn] n. 三聚氰胺
phosphoric acid 磷酸
soda ash ['səʊdəæʃ] 纯碱
sodium ['səʊdiəm] n. [化学] 钠
gluconate ['gluːkəʊneɪt] n. 葡（萄）糖酸盐（或酯）
hyposulfite [ˌhaɪpə'sʌlfaɪt] n. 次硫酸盐
quote [kwəʊt] n. 报价

With thanks and regards,
    Pharmachem Traders Pvt. Ltd.
      Sachin Pal
  103G，Block-F
  New Alipore
  Kolkata – 700 053
  Phone：033-2397-5003/5004
  Fax：033-2397-29789
  E-Mail：ptcal@cal. vsnl. net. in
  URL：www. nirkongrp. com

CIF Cost，Insurance and Freight 到岸价格
Kolkata［地名］（印度）加尔各答
FCL Full Container Load 全集装箱装载
item［ˈaɪtəm］n. 一件商品（或物品）
L/C Letter of Credit 信用证
acknowledge［əkˈnɒlɪdʒ］vt. 鸣谢

Comprehension

Choose the best answer according to the text.

1. According to the text，what is the main purpose of writing this letter? （　　）
 A. thanks for the last letter
 B. ask for more details of some certain chemicals
 C. try to build business relationships
 D. request a quote
2. The writer wants to （　　）.
 A. sale chemicals to the receiver
 B. buy chemicals from the receiver
 C. just get contact with the receiver
 D. get to know the price of the chemicals from the receiver
3. The letter is writing to （　　）.
 A. a buyer   B. a supplier   C. a manufacturer   D. a friend
4. Which statement is not true according to the text? （　　）
 A. 1) Price；2) COA and MSDS；3) Payment terms；4) Load port；5) The bank details；6) The name of companies in India to whom you have supplied in recent times should be mentioned in the replies.
 B. The price mentioned in the replies should include the price of the shipment and the insurance.
 C. They will pay the money as soon as they receive their orders.
 D. Pharmachem Traders Pvt. Ltd. has already do business in China for 20 years.

Write the formulae according to their English names

1. ammonium chloride   _____
2. acetic acid glacial   _____
3. sodium sulphide   _____
4. soda ash   _____
5. caustic soda   _____
6. melamine   _____

Practice

Please try to imitate this inquiry letter and write to Sinopec for some of its main products in the following blanks. Pay attention to the format.

Tips: Do you know the main products of Sinopec? If not, you'd better search in Sinopec's homepage.

_____
_____
_____
_____
_____
_____
_____
_____
_____
_____

Vocabulary building

Active words

appreciate
vt. 欣赏；感激

correspondence
n. 通信，信件

supply
vt. 供应，提供，供给
n. 供给，供应，提供；补给

minimize
vt. 把……减至最低数量［程度］

quote
vt. 引用；［商业］报价

item
n. 一些物品中的一项；项目

Exercise

Fill in each blank with a given word or expression in their right form.

appreciate   quote   CIF   minimize   supply   item

1. We _____ power to the three nearby towns.
2. While I love and _____ every single one of my friends, sometimes I wish they share some of my passions, so we could bond over them.
3. Emphasize differences between groups, but _____ differences between items within a group.
4. This way, we can collect everything we need to provide them with an accurate _____ without needing to have an actual conversation.
5. This is our offer for 10 000 pieces of table cloth at US _____ New York.
6. It's worth checking each _____ for obvious flaws.

Marketing & Selling

Extension

如何用英语介绍产品？

用英语介绍产品时，要根据所给信息介绍某种产品的品名、型号、原料、生产商、特点、功能以及使用说明和维护方法。特别注意以下几方面：

① 清楚地描述某产品的特征，包括产品的外观、颜色和体积等；

② 描述该产品的优点，让客户知道拥有它能带给自己的好处；

③ 用统计数字、众所周知的事实和别人的经验来证明某产品的优势，增强客户购买的信心。通常介绍时要提供该产品的图片。最好还要介绍产品在国内外的影响力，年销售量和出口量等。

The Model of Introducing the Chemical Product

It's my pleasure to introduce the product to you. We call the product _____（产品名称）which is characteristic of _____（产品特征）. Owing to its _____（产品优点），_____（产品名称）are well received in most _____（产品主要出口地区）countries. The annual output is _____（产品产量）. These items are most sellable in our market. _____（某公司的某产品）are not only as low-priced as other _____（同类产品），but they are distinctly superior in the following respects. Firstly, _____（产品最大优势）. Secondly, _____（产品第二大优势）and last but not least, _____（产品第三大优势）. To sum up, this product has excellent quality, reasonable price, distinctive features and a predictable large circulation.

Adapted from Shiyong Yingyu edited by Zi Wei

Practice

Paraffin wax, which is widely used in petroleum & chemical corporations, is one of the major products of CNPC. Please try to give a general introduction of the product showed in the following picture.

_____

_____

_____

_____

_____

_____

_____

Text C Sale & Purchase Contract for BRAZIL Iron Ore Fines

Contract Date:
Contract Number:
This contract ("Contract") is made and entered into by and between:
Seller:
Buyer:
Whereas, the buyer agrees to buy and the seller agrees to sell the below-mentioned goods for Asia, on the terms and conditions stated below.

CLAUSE 1: DEFINITION

In this contract, the following terms shall, unless otherwise specifically defined, have the following meanings:

① "Ore" means Iron Ore Fines of Brazil Origin.
② "U. S. Currency" means the currency of the United States of America freely transferable from and payable to an external account.
③ "Metric Tonne or MT" means a tonne equivalent to 1,000 kilogram.
④ "Wet basis" means Ore in its natural wet state.
⑤ "Dry basis" means Ore dried at 105 degrees Centigrade.
⑥ "DMT" means dry Metric Tonne

CLAUSE 2: COMMODITY

Ore.

CLAUSE 3: DELIVERY QUANTITY AND DELIVERY PERIOD

Quantity: _____ WMT (±10%) at Buyer's Option

Loading Port: Rio de Janeiro port, Brazil (at Seller's option)

Discharging Port: _____/_____ port, China (at Buyer's option)

Shipment: On or before _____, _____, 2015

CLAUSE 4: GUARANTEED SPECIFICATIONS

Chemical composition (on Dry basis)

Fe 64.50 % (rejection below 63.50%)
Al_2O_3 2.0% max
SiO_2 3.5% max

Words and expressions

purchase ['pɜːtʃes] v. 购买
contract ['kɒntrækt] n. 合同
ore [ɔː(r)] n. 矿
ore fine 矿粉
enter into 订立（协议）

whereas ['weər'æz] conj. 鉴于
term [tɜːm] n. 条款

clause [klɔːz] n. 条款
definition [ˌdefɪ'nɪʃn] n. 定义

currency ['kʌrənsi] n. 货币
transferable [træns'fɜːrəbl] adj. 可流通的
external [ɪk'stɜːnl] adj. 外部的
account [ə'kaʊnt] n. 账目
metric tonne ['metrɪktʌn] n. 公吨
equivalent [ɪ'kwɪvələnt] [化学] 当量的
kilogram ['kɪləɡræm] n. 公斤
wet basis 按湿量计算
dry basis 折干计算
commodity [kə'mɒdəti] n. 商品
quantity ['kwɒntəti] n. 量，数量
Rio de Janeiro ['riː(ː) əʊdədʒə'nɪərəʊ] 里约热内卢（巴西港市，州名）
at buyer's option 买方有权选择
at seller's option 卖方有权选择
discharging port 卸货港口
shipment ['ʃɪpmənt] n. 装运
guarantee [ˌɡærən'tiː] vt. 担保
specification [ˌspesɪfɪ'keɪʃn] n. 说明
rejection [rɪ'dʒekʃn] n. 拒绝
sulphur ['sʌlfə(r)] n. 硫磺
phosphorus ['fɒsfərəs] n. [化学] 磷

Sulphur　　0.01% max

Phosphorus　　0.06% max

Free moisture content loss at 105 degrees Centigrade shall be 8.00% max.

CLAUSE 5：PRICE

US $ per DMT CIF ＿＿＿＿ port，China (Incoterms 2000). The above price shall be based on 64.5% Fe (Rejection below Fe 63.5%) based on CIQ analysis report at discharging port and also rejection below Fe 64.0% based on loading port analysis.

Bank Address：

××××

CLAUSE 6：PRICE ADJUSTMENT

The prices of Ore stipulated in Clause 5 hereof shall be adjusted by the following bonuses and penalties：

ORE CONTENT

BONUS：For each 1.00% of Fe above 64.5%，the base price shall be increased by US $ 1.00 per DMT，fractions pro rata.

PENALTY：For each 1.00% of Fe below 64.5%，the base price shall be decreased by US $ 1.00 per DMT，fractions pro rata. But buyer has the right to reject Ore which is Fe content is below 63.5%.

OTHER ELEMENTS (IMPURITIES)：

If the composition of Ore in respect of Alumina (Al_2O_3), Silica (SiO_2), Sulphur (S) and Phosphorus (P) exceeds the respective guaranteed maximum as set forth in Clause 4 hereof，buyer shall accept such delivery of Ore by imposing penalties provided below，fractions pro rata.

① Al_2O_3

US $ 5 (Five) cents per DMT for each 1.00% in excess of 2.00%.

② SiO_2

US $ 5 (Five) cents per DMT for each 1.00% in excess of 3.50%.

③ S (Sulphur)

US $ 5 (Five) cents per DMT for each 0.01% in excess of 0.01%.

④ P (Phosphorus)

US $ 5 (Five) cents per DMT for each 0.01% in excess of 0.06%.

MOISTURE：

In the event that the free moisture loss at 105

free moisture content 游离水分含量

Incoterms 2000　2000年国际贸易术语解释通则
CIQ CHINA ENTRY-EXIT INSPECTION AND QUARANTINE BUREAU 中国出入境检验检疫局
stipulate ['stɪpjuleɪt] vt. 约定
hereof [,hɪər'ɒv] adv. 关于此点
adjust [ə'dʒʌst] v. 适应
bonus ['bəunəs] n. 奖金，额外津贴；红利
penalty ['penəlti] n. 惩罚；刑罚
fraction ['frækʃn] n. [数] 分数；一小部分，些微
pro rata [,prəu'rɑː tə] adv. 〈拉〉按比例，成比例
impurity [ɪm'pjuərəti] n. 污点；杂质
in respect of 关于
exceed [ɪk'siː d] vt. 超过；超越
respective [rɪ'spektɪv] adj. 各自的，分别的
impose [ɪm'pəuz] vt. 强加；征税
in excess of 超过，超出

degrees centigrade exceeds the respective guaranteed maximum as set forth in Clause 4 hereof, seller shall pay buyer half of the actual freight attributable to moisture content over 8% up to 9% including 9% and full actual freight attributable to moisture content over 9%.

CLAUSE 7: PAYMENT

Buyer shall open within seven (7) working days after the date of this contract, on at sight Letter of Credit ("L/C") in favor of seller providing for payment of the full invoice value of quantity of Ore. The L/C should contain the following terms and conditions:

① The L/C shall be issued or transferred by First Class Bank. All banking charges outside the L/C issuing bank including reimbursing charges and confirmation charges are for the account of the beneficiary.

② L/C shall allow for 10% more or less in value and quantity.

③ Charter party bills of lading acceptable.

④ Third party documents acceptable except for Draft and Invoice.

⑤ Partial shipment allowed.

⑥ Trans-shipment not allowed.

⑦ Spelling and other typographical errors are not considered as discrepancies.

(A) Provisional Payment

The aforesaid L/C shall be payable against seller's sight draft (s) for the amount of ninety-five (95) percent of the CIF value of the shipment accompanied by the documents as stipulated in Clause 8 hereof. The Certificate of Weight issued by SGS Brazil Private, Ltd. ("SGS Brazil") in Brazil by survey of ship's draft together with the Certificate of Analysis of sample and of the percentage of the free moisture loss at 105 degrees centigrade issued by SGS Brazil shall be a basis for the seller's provisional invoice for the provisional payment.

(B) Final Payment

The balance (+/−) of the CIF value of the shipment after the provisional payment to seller shall be settled in accordance with seller's draft payable at sight together with the final invoice and the documents stated in Clause 8 hereof or by

freight [freit] n. 货运；运费
attributable [əˈtrɪbjətəbl] adj. 可归因于……的；由……引起的

in favor of（支票）以某人/某部门为收款人
invoice [ˈɪnvɔɪs] n. 发票；发货单
issue [ˈɪʃuː] vt. 发行
banking charges 银行手续费
reimbursing charges 偿付费
beneficiary [ˌbenɪˈfɪʃəri] n. 受益人

charter [ˈtʃɑːtə(r)] n. 许可证
bills of lading 提货单

typographical [ˌtaɪpəˈɡræfɪkl] adj. 印刷上的
discrepancy [dɪsˈkrepənsi] n. 矛盾
provisional [prəˈvɪʒənl] adj. 暂时的
aforesaid [əˈfɔːsed] adj. 上述的，前述的（常用于法律文件）

telegraphic transfer by Seller to Buyer in accordance with Buyer's debit note to Seller together with the documents (except Seller's final invoice) stated in Clause 8 hereof within ten (10) days from the date of receipt of the said documents, as the case may be. The validity of the L/C should be maintained in accordance with the above.

Seller's final invoice or Buyer's debit note for the settlement of the balance of the CIF value of the shipment shall be based on the certificate issued by the State General Administration of the People's Republic of China for Quality Supervision and Inspection and Quarantine or its branches (collectively "CIQ China") as provided in Clause 8 and Clause 9 hereof. If an umpire analysis is required under the Clause 10 hereof, payment adjustment arising from this will be made when the umpire's certificate is available.

If Buyer is not able to submit to Seller the Inspection Certificate issued by CIQ China as per Clause 9 hereof within seventy five (75) days after the completion of discharging, the Draft Survey Certificate of Weight and Certificate of Analysis at loading port shall be the base of the final invoice or debit note.

If the cargo is rejected based on CIQ analysis report seller shall refund the amount vide Clause 7 (A) Provisional Payment, within 30 days after the receipt of the CIQ analysis report.

Buyer's banking details:
Seller's bank details:

telegraphic transfer 电汇
debit note 借方通知，收款单

umpire analysis 仲裁分析

submit [səb'mɪt] vi. 顺从，服从；提交
discharging [dɪs'tʃɑːdʒɪŋ] n. 卸料
loading port 装货港
cargo ['kɑːɡəʊ] n. 货物；负荷
vide ['viːdeɪ] v. 请见，参阅

Comprehension

Fill in the form according to the text.

CLAUSE IN A CONTRACT	DETAILS
DEFINITION	
COMMODITY	
DELIVERY QUANTITY AND DELIVERY PERIOD	
GUARANTEED SPECIFICATIONS	
PRICE	
PRICE ADJUSTMENT	
PAYMENT	

Vocabulary building

Active words

purchase
v. 购买
n. 购买行为

contract
n. 合同
v. 签合同

currency
n. 货币

equivalent
n. 等值物；对应物
adj. 相当的，等效的

guarantee
n. 保证

v. 承诺做

specification
n. 规格；具体要求

reject
v. 拒绝

adjust
v. 改变（行为或观点）以适应

submit
v. 屈服；投降；被迫接受；提交，递呈（建议、报告或请求）

Useful expressions

enter into
加入（讨论）；订立（协议）；开始（关系）；成为……的一部分，成为……的一个因素；构成

Exercise

Fill in each blank with a given word or expression in their right form.

specification   adjust   guarantee   contract   currency   submit   equivalent   enter into

1. The minority should _____ majority.
2. She will have to _____ herself to new conditions.
3. Have you get a product of this _____?
4. Words of mouth being no _____, a written statement is hereby given.
5. Eight kilometers is roughly _____ to five miles.
6. Many countries charge departure tax in US dollars rather than local _____.
7. China will not _____ alliance with any big power.
8. They can transfer or share the _____ with whoseever they choose.

Extension

Dialogues 1

John is working in the marketing department in a chemical plant. In this dialogue he is showing Vincent, the foreign client, around the chemical plant. They are talking about one of the products—paraffin.

John： Let me take you to our showroom to look at our latest product.

Vincent:	It's very kind of you. I am interested in your product, paraffin. Would you like to introduce it in details for me?
John:	It's my pleasure. Would you like to have a look at our samples first?
Vincent:	Yes, thank you. I want to have a clearer picture of the product.
John:	As our products have all the features you need and are 20% cheaper compared with that Japanese brand, I strongly recommend them to you.
Vincent:	What are the major types do you have?
John:	We have 54#, 56#, 58#, 60#, 62# and 64#. Our products are the bestsellers of the similar kinds, and it wins high reputation among our customers.
Vincent:	Do you check the package?
John:	Sure, but we've recently started to cooperate with packing companies, too.
Vincent:	How about the common color?
John:	White. Do help yourself to a brochure, sir.
Vincent:	I've studies all your brochures. Could you give us some detailed information of your products?
John:	The catalogue contains the illustrations and description of the large variety of our supplies.
Vincent:	How about the output?
John:	Although the output is very large, there's been a big rush for it lately.
Vincent:	How much is the 58# paraffin? If your price are favorable, I will place the order right away.
John:	¥15000 per ton. As a special sign of encouragement, we'll consider accepting a dicount of 5% at this sales purchasing stage.
Vincent:	Thank you for your introduction.
John:	You are welcome. I'm sure that you will find our product is really competitive in the market. Then I'll give you a tour of our office.

Dialogues 2

The negotiation between John and Vincent finally comes to an end. They are now signing a sale & purchase contract for paraffin 58#.

John:	The contract is ready. Will you please check up the particulars and see if everything is in order?
Vincent:	Well, everything is all right. There's only one thing I would like to point out that is timely delivery. You know our customers are in urgent need of the goods. If you fail to deliver the goods at the time stipulated in the contract, they may turn elsewhere for substitution. In that case, we just can't stand the loss.
John:	You are assured that the shipment will be duly delivered; we must have your L/C at least one month before the time of shipment.
Vincent:	Certainly. When I get back, I'll open an L/C for the whole quantity as soon as possible.
John:	Good. Another thing, the stipulations in the relevant credit should strictly confirm to the terms stated in the contract in order to avoid subsequent

	amendment. If that does happen, shipment will possibly be delayed.
Vincent:	All right. I'll see to it.
John:	Any other question?
Vincent:	No, nothing more.
John:	Shall we sign the contract now?
Vincent:	With pleasure.
John:	Now, please countersign it. You may keep one original and two copies for yourself.
Vincent:	Thank you.
John:	I'm glad our negotiation has come to a successful conclusion. I hope this mark the beginning of long and stable business relations between us.
Vincent:	I hope so, too.

Practice

Try to act these two dialogues out in pairs, and cross out what information the client concerns most.

Reading material

With the progress of the Chinese oil industry, there are three main players in this area. This article aims to give some general ideas about them.

Sinopec (see Fig. 6.1)

China Petrochemical Corporation (Sinopec Group) is a super-large petroleum and petrochemical enterprise group established in July 1998 on the basis of the former China Petrochemical Corporation. Sinopec Group is a state-owned company solely invested by the State, functioning as a state-authorized investment organization in which the state holds the controlling share. Headquartered in Beijing, Sinopec Group has a registered capital of RMB 182 billion.

Sinopec Group's key business activities include: industrial investment and investment management; the exploration, production, storage and transportation (including pipeline transportation), marketing and comprehensive utilization of oil and natural gas; oil refining; the wholesale of gasoline, kerosene and diesel; the production, marketing, storage, transportation of petrochemicals and other chemical products; the design, construction and installation of petroleum and petrochemical engineering projects; the overhaul and maintenance of petroleum and petrochemical equipment; the manufacturing of electrical and mechanical equipment; the research, development, application and consulting services of technology, information and alternative energy products, the import and export of commodities and technologies both for the Group and as a proxy (with the

Fig. 6.1 Sinopec

exception of those commodities and technologies that are either banned by the State or to be carried out by the state-designated companies).

CNPC(see Fig. 6.2)

China National Petroleum Corporation (CNPC) is an integrated international energy company.

Fig. 6.2 CNPC

Based in China, they have oil and gas assets and interests in 37 countries in Africa, Central Asia-Russia, America, the Middle East, Asia-Pacific, and other regions.

CNPC is China's largest oil and gas producer and supplier, as well as one of the world's major oilfield service providers and a globally reputed contractor in engineering construction, with businesses covering petroleum exploration & production, natural gas & pipelines, refining & marketing, oilfield services, engineering construction, petroleum equipment manufacturing and new energy development, as well as capital management, finance and insurance services.

Crude output: 112.60 million tons per year in China.

Natural gas output: 88.84 billion cubic meters per year in China.

Crude oil production: 54% of China's total.

Natural gas production: 75% of China's total.

Crude runs: 146.02 million tons per year in China.

Domestic service stations: 20,272(1).

Domestic pipelines: 72,878 kilometers, including 17,640 kilometers for crude oil (70% of China's total), 45,704 kilometers for natural gas (80% of China's total) and 9,534 kilometers for refined products (47% of China's total).

Providing oilfield services and engineering construction in 67 countries around the world.

CNOOC(see Fig. 6.3)

China National Offshore Oil Corporation ("CNOOC"), the largest offshore oil & gas producer in China, is a state-owned company operating directly under the State-owned Assets Supervision and Administration Commission of the State Council of the People's Republic of China. Headquartered in Beijing, founded in 1982, CNOOC has evolved from an upstream oil & gas company to an international energy company with promising core businesses and a complete industrial chain.

Fig. 6.3 CNOOC

Now, CNOOC businesses cover the main segments of oil & gas exploration and development, engineering & technical services, refining and marketing, natural gas and power generation, and financial services.

In 2013, the company's overseas businesses covered more than 40 countries and regions, with overseas assets proportion up to 40%, overseas revenue proportion up to 30% and localization rate of overseas employees up to 82%.

In 2014, CNOOC's ranking in the Fortune Global 500 rose to 79, which was 14 higher than that in 2013. Standard & Poor's and Moody's continue to rate the company with credit ratings of AA-and Aa3, the highest for a Chinese corporation.

Keeping in line with the Second Leap Forward Development Program, CNOOC will continue to focus on the transformation of economic development pattern, the industry structural adjustment and the improvement of development quality and efficiency.

Adapted from ir2. mofcom. gov. cn/

Work in groups

1. Discuss with your group members about rules of word-formation of following words, and write their corresponding verb forms.

corporation		organition	
investment		management	
utilization		transpotation	

2. Discuss with your group members about the meaning of the following phrases.

Crude output		Crude runs	
Natural gas output		Domestic service stations	
Crude oil production		Domestic pipelines	
oilfield services		engineering construction	

3. Discuss with your group members about the meaning of the following sentences.

1. Headquartered in Beijing, founded in 1982, CNOOC has evolved from an upstream oil & gas company to an international energy company with promising core businesses and a complete industrial chain.

_____

2. Now, CNOOC businesses cover the main segments of oil & gas exploration and development, engineering & technical services, refining and marketing, natural gas and power generation, and financial services.

_____

3. In 2014, CNOOC's ranking in the Fortune Global 500 rose to 79, which was 14 higher than that in 2013.

_____

Supplementary knowledge

Samples of Some International Trade Documents

COMMERCIAL INVOICE

ISSUER: GREAT WALL TRADING CO., LTD. RM201, HUASHENG BUILDING, NINGBO, P. R. CHINA			
TO: F. T. C. CO. AKEKSANTERINK AUTO P. O. BOX 9, FINLAND	NO. GW2005M06-2		DATE 22 MAY, 2005
TRANSPORT DETAILS FROM NINGBO TO HELSINKI BY SEA PARTIAL SHIPMENT: NOT ALLOWED TRANSHIPMENT: ALLOWED SHIPPMENT AT THE LATEST MAY 30, 2005	S/C NO. GW2005M06		L/C NO. LRT9802457
	TERMS OF PAYMENT L/C AT SIGHT		

Marks and Numbers	Number and kind of package Description of goods	Quantity	Unit Price	Amount
ROYAL 05AR225031 JEDDAH C/N: 1-UP	P. P INJECTION CASES ZL0322+BC05 230SETS ZL0319+BC01 230SETS DETALS AS PER SALES CONTRACT GW2005M06 DATED APR.22,2005 CIF HESINKI			
	ZL0322+BC05 ZL0319+BC01	230CTNS 230CTNS	USD42.00 USD41.00	USD9660.00 USD9430.00
	TOTAL:	460CTNS	USD83.00	USD19090.00

SAY TOTAL: NINTEEN THOUSAND AND NINTY ONLY

THE NAME AND ADDRESS OF THE MANUFACTURER: GREAT WALL TRADING CO., LTD.

SIGNATURE: GREAT WALL TRADING CO., LTD.
RM201, HUASHENG BUILDING,
NINGBO, P. R. CHINA

SIGNED COMMERCIAL INVOICE 1 ORIGINAL AND 5 COPIES

1. Shipper Insert Name, Address and Phone		B/L No.
GREAT WALL TRADING CO., LTD. RM201, HUASHENG BUILDING, NINGBO, P. R CHINA		CSC020867

 中远集装箱运输有限公司
COSCO CONTAINER LINES

TLX: 33057 COSCO CN
FAX: +86(021) 6545 8984

ORIGINAL
Port-to-Port or Combined Transport

BILL OF LADING

2. Consignee Insert Name, Address and Phone
TO ORDER

| 3. Notify Party Insert Name, Address and Phone |
| (It is agreed that no responsibility shall attach to the Carrier or his agents for failure to notify) |
| F. T. C. CO.
AKEKSANTERINK AUTO P. O. BOX 9, FINLAND |

RECEIVED in external apparent good order and condition except as otherwise noted. The total number of packages or unites stuffed in the container, The description of the goods and the weights shown in this Bill of Lading are Furnished by the Merchants, and which the carrier has no reasonable means Of checking and is not a part of this Bill of Lading contract. The carrier has Issued the number of Bills of Lading stated below, all of this tenor and date, One of the original Bills of Lading must be surrendered and endorsed or signed against the delivery of the shipment and whereupon any other original Bills of Lading shall be void. The Merchants agree to be bound by the terms And conditions of this Bill of Lading as if each had personally signed this Bill of Lading.
SEE clause 4 on the back of this Bill of Lading (Terms continued on the back Hereof, please read carefully).
*Applicable Only When Document Used as a Combined Transport Bill of Lading.

4. Combined Transport * Pre - carriage by	5. Combined Transport* Place of Receipt
6. Ocean Vessel Voy. No. YANGFNA V.009W	7. Port of Loading NINGBO
8. Port of Discharge HELSINKI	9. Combined Transport * Place of Delivery

Marks & Nos. Container / Seal No.	No. of Containers or Packages	Description of Goods (If Dangerous Goods, See Clause 20)	Gross Weight Kgs	Measurement
ROYAL 05AR225031 JEDDAH C/N: 1-UP	CBHU 0611758/ 25783 CY/CY PACKED IN 460CTNS	P. P INJECTION CASES ZL0322+BC05 230SETSZL0319+BC01 230SETS DETALS AS PER SALES CONTRACT GW2005M06 DATED APR.22,2005CIF HESINKI L/C NO. LRT9802457 DATE. APRIL 28,2005 CY/CY CONTAINER NO.********* ZL0322+BC05		
			4255KGS	34M3
		ZL0322+BC05	4255KGS	34M3
TOTAL	460CTNS	ZL0319+BC01	8510KGS	68M3
		FREIGHT PREPAID		
		Description of Contents for Shipper's Use Only (Not part of This B/L Contract)		

10. Total Number of containers and/or packages (in words)
    Subject to Clause 7 Limitation

11. Freight & Charges Declared Value Charge	Revenue Tons	Rate	Per	Prepaid V	Collect
Ex. Rate:	Prepaid at CHINA	Payable at		Place and date of issue MAY 25, 2005 .NINGBO, P. R CHINA.	
	Total Prepaid	No. of Original B(s)/L THREE		Signed for the Carrier, COSCO CONTAINER LINES ANDYLVKING	

LADEN ON BOARD THE VESSEL
DATE MAY 25, 2005 BY COSCO CONTAINER LINES
ENDORSED IN BLANK ON THE BACK

1.Exporter GREAT WALL TRADING CO., LTD. RM201, HUASHENG BUILDING, NINGBO, P. R CHINA	Certificate No. # CERTIFICATE OF ORIGIN ## OF # THE PEOPLE'S REPUBLIC OF # CHINA
2.Consignee F. T. C. CO. AKEKSANTERINK AUTO P. O. BOX 9, FINLAND	
3.Means of transport and route FROM NINGBO,P.R CHINA TO HELSINKI BY SEA	5.For certifying authority use only
4.Country / region of destination HELSINKI	

6.Marks and numbers	7.Number and kind of packages; description of goods	8.H.S.Code	9.Quantity	10.Number and date of Invoices
ROYAL 05AR225031 JEDDAH C/N: 1-UP	P. P INJECTION CASES ZL0322+BC05 230SETS ZL0319+BC01 230SETS DETALS AS PER SALES CONTRACT GW2005M06 DATED APR.22,2005 CIF HESINKI	230CTNS 230CTNS	230CTNS 230CTNS	GW2005M06-2 MAY 22, 2005

11.Declaration by the exporter
  The undersigned hereby declares that the above details and statements are correct, that all the goods were produced in China and that they comply with the Rules of Origin of the People's Republic of China.
 	
GREAT WALL TRADING CO., LTD.
RM201, HUASHENG BUILDING, NINGBO, P. R CHINA
 	
NINGBO CHINA ,MAY 20,2005
--- 	
Place and date, signature and stamp of authorized signatory
IN 2 COPIES
	12.Certification
	  It is hereby certified that the declaration by the exporter is correct.
	
	ANDYLVKING
	--- 	
	Place and date, signature and stamp of certifying authority

Exercise

Read the sample documents and answer the questions.

1. Do you know the Chinese names of these 3 documents?

2. In the first invoice, who was the seller? Who was the buyer? And what were the goods?

3. In the second bill of lading, where did the goods load? What was the destination of the goods?

4. In the third document, what is the meaning of the term "consignee"?

5. What is the function of the third document?